# Mobile App Reverse Engineering

Get started with discovering, analyzing, and
exploring the internals of Android and iOS apps

**Abhinav Mishra**

BIRMINGHAM—MUMBAI

# Mobile App Reverse Engineering

Copyright © 2022 Packt Publishing

**Group Product Manager**: Vijin Boricha
**Associate Publishing Product Manager**: Prachi Sawant
**Senior Editor**: Athikho Sapuni Rishana
**Content Development Editor**: Sayali Pingale
**Technical Editor**: Nithik Cheruvakodan
**Copy Editor**: Safis Editing
**Associate Project Manager**: Neil Dmello
**Proofreader**: Safis Editing
**Indexer**: Pratik Shirodkar
**Production Designer**: Prashant Ghare
**Marketing Co-ordinator**: Hemangi Lotlikar

First published: April 2022
Production reference: 1200422

Published by Packt Publishing Ltd.
Livery Place
35 Livery Street
Birmingham
B3 2PB, UK.

978-1-80107-339-4

www.packt.com

*Dedicated to the late Rajendra Singh, a passionate teacher, accomplished author, and learner who inspired many, like me, to seek knowledge above everything else. To my wife, Kanika Singh, for being the support I have always needed. And to my mother and father, for being role models to me and for motivating me.*

# Contributors

## About the author

**Abhinav Mishra** is the founder and director of Enciphers, a cybersecurity consulting and training company. Abhinav has over a decade of extensive experience in finding and fixing security issues in web, mobile, and infrastructure applications. He has performed penetration tests on more than 500 mobile applications and has discovered thousands of critical vulnerabilities.

Abhinav completed his engineering degree in 2011 and since then has been involved in penetration testing and securing applications and infrastructure. Prior to founding Enciphers, Abhinav worked with Fortune 500 and giant tech companies as part of their security teams. In his spare time, he is a traveler, adventure seeker, and drone hobbyist.

*I would like to thank Manoj Jain, a skilled Android developer with almost a decade of experience, and Mohammad Haroon, who is a passionate iOS developer with 12+ years of experience in developing Swift/Objective and C/C++/C apps. Both of them assisted in developing the SecureStorage app, used in the book. Their contribution to this book is highly appreciated.*

# About the reviewer

**Anant Shrivastava** is the founder of a research firm named Cyfinoid Research. His last job was as a technical director for NotSoSecure Global Services. He has been active in the Android security field since the early days of Android development (2011). He has been a trainer and speaker at various international conferences (Black Hat – USA, Asia, EU, Nullcon, c0c0n, and many more). Anant also leads the open source projects Android Tamer and Code Vigilant. He also maintains the archive portal named Hacking Archives of India. In his spare time, he likes to take part in open communities geared to spreading information security knowledge, including the null community, Garage4hackers, Hasgeek, and OWASP.

> *I truly believe all of us in the technical world are standing on the shoulders of giants. The giants for me are the open communities, such as null, Garage4hackers, Hasgeek, and OWASP, where access to information is unrestricted and people are interested in helping one another. I am deeply indebted to all the communities and the people running these communities. I am also thankful to my whole family for providing all the support and tolerating my busy schedule and still standing by my side. I would also like to do a special shout-out to my son, Aarush, whose smile gives me a reason to keep going.*

# Table of Contents

# Section 2: Mobile Application Reverse Engineering Methodology and Approach

## 3

## Reverse Engineering an Android Application

## 4

## Reverse Engineering an iOS Application

## 5

## Reverse Engineering an iOS Application (Developed Using Swift)

# Section 3: Automating Some Parts of the Reverse Engineering Process

## 6

## Open Source and Commercial Reverse Engineering Tools

## 7

## Automating the Reverse Engineering Process

## 8

## Conclusion

## Index

## Other Books You May Enjoy

# Preface

Mobile application reverse engineering is an important skill for penetration testers, malware analysts, and application security professionals in general. This book talks about how Android and iOS applications are developed, how to reverse engineer them, different case studies of security issues discovered through reverse engineering, and how to automate the reverse engineering and analysis part.

The book helps in understanding the internals of modern Android and iOS apps and how you can reverse engineer application packages (APK and IPA). Here, you can start your journey of creating a reverse engineering mobile application.

## Who this book is for

This book is for cybersecurity professionals, security analysts, mobile application security enthusiasts, and penetration testers interested in understanding the internals of iOS and Android apps through reverse engineering. Basic knowledge of reverse engineering as well as an understanding of mobile operating systems such as iOS and Android and how mobile applications work on them are required.

## What this book covers

*Chapter 1, Basics of Reverse Engineering – Understanding the Structure of Mobile Apps,* talks about the reverse engineering fundamentals, common terminologies, and Android and iOS application fundamentals.

*Chapter 2, Setting Up a Mobile App Reverse Engineering Environment Using Modern Tools,* gets you familiar with the tools used in the reverse engineering of mobile (Android and iOS) applications, and then sets up an environment for reverse engineering by installing the same tools in a virtual machine. The chapter also mentions Mobexler, a mobile application penetration-testing platform.

*Chapter 3, Reverse Engineering an Android Application,* deep-dives into how Android apps are developed, their internal components, structure, format, and binary details, and finally, how to reverse an Android application package to extract the Java as and `smali` code.

*Chapter 4, Reverse Engineering an iOS Application*, discusses how iOS apps are developed, understanding the iOS executable format, exploring more iOS app reverse engineering tools and their usage, and finally, reverse engineering an iOS application package.

*Chapter 5, Reverse Engineering an iOS Application (Developed Using Swift)*, details the difference between Objective-C and Swift applications from a developer's perspective and also explains the process of reverse engineering a Swift application using the Radare2 reverse engineering tool.

*Chapter 6, Open Source and Commercial Reverse Engineering Tools*, discusses some common open source as well as commercial (closed source) tools for reverse engineering, together with real-world case studies for reverse engineering and the required capabilities in a reverse engineering tool.

*Chapter 7, Automating the Reverse Engineering Process*, explains when it might be a good idea to automate some parts of reverse engineering, and how to do that. This chapter also looks at some case studies to explain how automation can be performed for some test cases.

*Chapter 8, Conclusion*, talks about what to do next, and what other skills might be good to learn if you want to continue this journey of reverse engineering.

# To get the most out of this book

| Software/hardware covered in the book | OS requirements |
| --- | --- |
| The basics of Linux (Ubuntu) | Linux (Ubuntu) |
| Virtualization tools | VirtualBox and VMware |
| The basics of shell scripting | Linux |

# What not to expect from the book

The book is about getting started with mobile application reverse engineering. It should not be treated as a book to become a reverse engineering expert. The book specifically talks about how to reverse engineer mobile apps; the basics of binary reversing is not really covered in the book.

# Download the color images

We also provide a PDF file that has color images of the screenshots/diagrams used in this book. You can download it here: `https://static.packt-cdn.com/downloads/9781801073394_ColorImages.pdf`.

# Conventions used

There are a number of text conventions used throughout this book.

`Code in text`: This indicates code words in text, database table names, folder names, filenames, file extensions, pathnames, dummy URLs, user input, and Twitter handles. Here is an example: "To analyze the function in visual mode or a more visual presentation, we can use the visual mode, using the `VV` command."

A block of code is set as follows:

```
-(void)loginInAppUsing :(NSString*) userName and : (NSString*)
password{

    NSString* str1 = userName;
    NSString* str2 = password;

```

Any command-line input or output is written as follows:

```
# ./ghidraRun
```

**Bold**: This indicates a new term, an important word, or words that you see onscreen – for example, words in menus or dialog boxes appear in the text like this. Here is an example: "This will open the Ghidra project window. We can choose the **Tools | CodeBrowser** option to open the code browser utility."

> **Tips or Important Notes**
> Appear like this.

# Disclaimer

*The information within this book is intended to be used only in an ethical manner. Do not use any information from the book if you do not have written permission from the owner of the equipment. If you perform illegal actions, you are likely to be arrested and prosecuted to the full extent of the law. Neither Packt Publishing nor the author of this book takes any responsibility if you misuse any of the information contained within the book. The information herein must only be used while testing environments with proper written authorization from the appropriate persons responsible.*

# Get in touch

Feedback from our readers is always welcome.

**General feedback**: If you have questions about any aspect of this book, mention the book title in the subject of your message and email us at customercare@packtpub.com.

**Errata**: Although we have taken every care to ensure the accuracy of our content, mistakes do happen. If you have found a mistake in this book, we would be grateful if you would report this to us. Please visit www.packtpub.com/support/errata, selecting your book, clicking on the Errata Submission Form link, and entering the details.

**Piracy**: If you come across any illegal copies of our works in any form on the internet, we would be grateful if you would provide us with the location address or website name. Please contact us at copyright@packt.com with a link to the material.

**If you are interested in becoming an author**: If there is a topic that you have expertise in and you are interested in either writing or contributing to a book, please visit authors.packtpub.com.

# Share Your Thoughts

Once you've read *Mobile App Reverse Engineering*, we'd love to hear your thoughts! Scan the QR code below to go straight to the Amazon review page for this book and share your feedback.

https://packt.link/r/1801073392

Your review is important to us and the tech community and will help us make sure we're delivering excellent quality content.

# Section 1: Basics of Mobile App Reverse Engineering, Common Tools and Techniques, and Setting up the Environment

This section explains from scratch the reverse engineering fundamentals, terminologies, the tools used, and setting up an environment using these tools. The chapters will also explain some basic uses for those tools and the structure of mobile apps.

This part of the book comprises the following chapters:

- *Chapter 1, Basics of Reverse Engineering – Understanding the Structure of Mobile Apps*
- *Chapter 2, Setting Up a Mobile App Reverse Engineering Environment Using Modern Tools*

# 1

# Basics of Reverse Engineering – Understanding the Structure of Mobile Apps

All of us use cell phones in our daily lives now, and their usage has grown to such a crucial level that people frequently name *cell phones* as one of the *three things you can't live without*, after food and water. Cell phones handle almost every task, from managing funds in bank accounts and investments to travel bookings, shopping, and health appointments.

To perform these tasks, cell phones have mobile apps. These apps handle a majority of your data and help you perform tasks.

As these modern mobile apps handle sensitive user information, perform critical tasks, and provide access to a huge array of resources on the internet, the security of the data being handled and the operations performed on it also need to be improved.

A mobile application *penetration tester* tests the security of mobile applications in order to find vulnerabilities. To find the vulnerabilities, the tester is required to understand the internal working and logics of the application. These details can be found in the source code of the application. However, the penetration testers do not always have the source code to hand, as in the case of a black-box penetration test. During a black-box penetration test, all that the penetration tester has is the application package, that is, the **Android Application Package (APK)** or **iOS App Store Package (IPA)** file. In such a case, to understand the working of the app, they need to unpack the application package and get the source code.

Reverse engineering is the technique of dismantling an object to study its internal designs, code, logic, and so on. Reverse engineering mobile applications is the process of disassembling/dismantling an app to reveal its code and internal logic, components, and so on.

In this chapter, we're going to cover the following main topics:

- Reverse engineering fundamentals
- Android application fundamentals
- iOS application fundamentals

We will learn about the basics of reverse engineering and how mobile applications are built. These fundamentals are important to understand before we can jump into the actual task of reverse engineering modern apps.

# Technical requirements

Android Studio and Xcode are required to complete the relevant hands-on exercises. Xcode is Apple's **integrated development environment (IDE)** for macOS, used to develop software for macOS, iOS, iPadOS, watchOS, and tvOS. Android Studio is the official IDE for Google's Android operating system. An Apple laptop/desktop (Mac) can install and run both Xcode and Android Studio, whereas other laptops/desktops running Windows or Linux will only be able to support Android Studio.

For more information, please refer to the following links:

- Android Studio: `https://developer.android.com/studio`
- Xcode: `https://developer.apple.com/xcode/`

# Reverse engineering fundamentals

Let's first understand the fundamentals of reverse engineering, why it is needed, and what steps are involved.

As mentioned earlier in this chapter, reverse engineering is the technique of dismantling an object to study its internal designs, code, and logic.

When a developer builds a mobile app, they choose a programming language (according to the targeted platform – Android, iOS, or both), write the code for the functionalities they want, and add resources such as images, certificates, and so on. Then the code is compiled to create the application package.

While reverse engineering the same app, the reverse engineer dismantles the application package to the components and code.

Some of the frequently used terms in reverse engineering are the following:

- **Decompilation**: This is the process of translating a file from a low-level language to a higher level language. The tool used to perform decompilation is called a **decompiler**. A decompiler takes a binary program file and changes this program into a higher-level structured language. The following diagram illustrates the decompilation process:

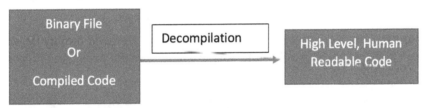

Figure 1.1 – Diagram of the decompilation process

- **Disassembling**: This is the process of transforming machine code (in an object code binary file) into a human-readable mnemonic representation called **assembly language**. The tool used to perform disassembly is called a **disassembler** as it does the opposite of what an assembler does. The following diagram illustrates the disassembly process:

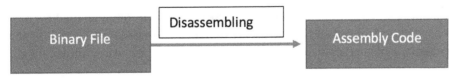

Figure 1.2 – Diagram of the disassembly process

A simple binary disassembled in a disassembling tool, Hopper, looks as follows:

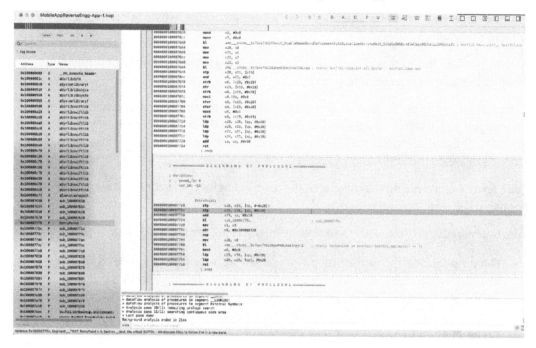

Figure 1.3 – Disassembled binary in Hopper

- **Debugging**: This is a technique that allows the user to view and modify the state of a program at runtime. The following diagram illustrates the debugging process:

Figure 1.4 – Diagram of the debugging process

Understanding the different methodologies and approaches used in reverse engineering is very important. We will be using all these concepts in further chapters of this book.

Now that we have seen the fundamentals of reverse engineering, let's explore how mobile applications, that is, Android and iOS apps, are developed. We will now be looking into the components, structure, and concepts behind the mobile application fundamentals.

# Android application fundamentals

Native Android applications are written mainly in Java or Kotlin. The Android SDK tools compile the code along with any data and resource files into an APK or an Android App Bundle. The compiled application is in a specific format, specified by the extension .apk. That is, an Android package is an archive file containing multiple application files and metadata.

---

**Fun Fact**

Rename the file extension of an APK to .zip and use unzip to open. You will be able to see its contents.

---

The following are the major components of an APK:

- AndroidManifest.xml: The application manifest file containing app details such as the name, version, referenced libraries, and component details in XML format. The Android operating system relies on the presence of this file to identify relevant information about the application and related files.

- Dalvik executable files (classes.dex files).

- META-INF:

  - MANIFEST.MF (manifest file)

  - CERT.RSA (certificate of the application)

  - CERT.SF (list of resources with SHA-1 digest of the corresponding lines in the MANIFEST.MF file)

- lib: This contains the compiled code that is specific to a selection of processors, as follows:

  - armeabi: Compiled code for all ARM-based processors

  - armeabi-v7a: Compiled code for all processors based on ARMv7 and above

  - x86: Compiled code for x86 processors

  - mips: Compiled code for MIPS processors

- res: Resources that are not compiled into resources.arsc.

- assets: Contains application assets.

- resources.arsc: Pre-compiled resources.

> **Important Note**
>
> Java code in Android devices does not run in the **Java Virtual Machine** (**JVM**). Rather, it is compiled in the **Dalvik Executable** (**DEX**) bytecode format. A DEX file contains code that is ultimately executed by Android Runtime.

Let's see how to create a simple *hello world* application for Android and then unzip it to look at its components:

1. Android apps are developed using Android Studio. Download and install the latest version of Android Studio from `https://developer.android.com/studio`:

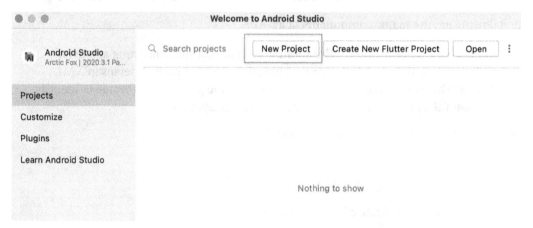

Figure 1.5 – Creating a new project in Android Studio

2. Let's choose the **New Project** option and select the **Empty Activity** option:

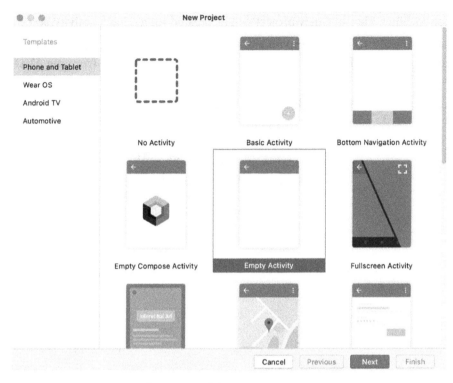

Figure 1.6 – Selecting project type

3. On the next screen, fill in all the details as shown in the following screenshot. You can choose the name as you please:

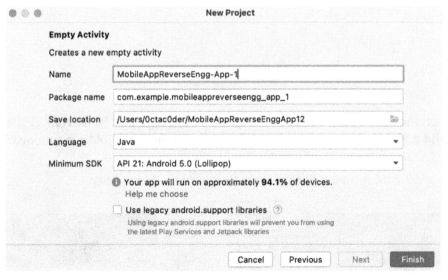

Figure 1.7 – Project details

4.  Once you click **Finish**, a new project will be created for a default activity/screen app.

5.  You can now try to run the app on any attached Android device, or the virtual Android emulator. For the latter, create a virtual Android device from the **AVD** menu.

6.  Once the app runs successfully, we will try to extract the application package for this app from Android Studio:

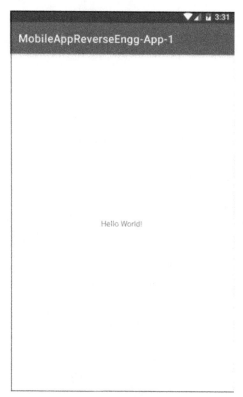

Figure 1.8 – Running the app on the emulator

7.  To get the APK from Android Studio, go to the **Build | Build Bundle(s)/APK(s) | Build APK(s)** menu option. Once generated, navigate to the folder mentioned in the **Locate** option and copy the APK.

8.  Once the APK is copied, change the extension of the file to `.zip`:

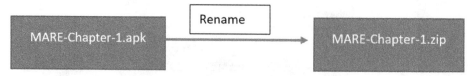

Figure 1.9 – Diagram of rename process

9.  Use any archive tool to unzip the file and extract its contents:

```
# unzip MARE-Chapter-1.zip
```

For reference, the result is as follows:

Figure 1.10 – Extracting the content of the APK, after renaming it to .zip

10. Let's analyze the components inside the APK and compare it with the list here (*Android application fundamentals*):

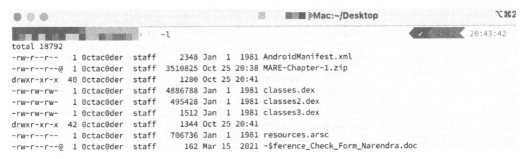

Figure 1.11 – Extracted content of the APK

The following diagram shows the processes of forward and reverse engineering an Android application:

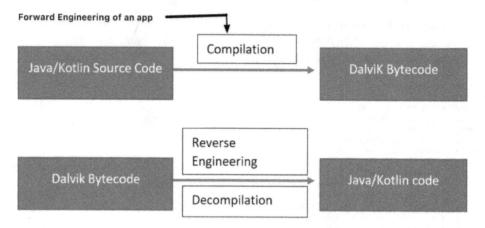

Figure 1.12 – The forward and reverse engineering processes with an Android application

Android applications are mainly developed using Java and Kotlin. The internals of an Android package are the same whether it is based on Java or Kotlin. Therefore, the approach to reverse engineer the application is also the same.

We've now learned about the fundamentals of Android applications. iOS apps are also packaged into a specific format and have a specific structure. Let's look into the iOS application fundamentals now.

# iOS application fundamentals

Similar to Android, iOS applications also come in a specific zipped format called **IPA**, or an **iOS App Store Package**. iOS application packages can also be renamed by changing the extension to ZIP and then the components can be extracted, though the components of an iOS application package differ from those of an Android one.

iOS apps are mainly built using Objective-C and Swift, both of which can be disassembled using a disassembler such as Hopper or Ghidra. In Objective-C applications, methods are called via dynamic function pointers, which are resolved by name during runtime. These names are stored intact in the binary, making the disassembled code more readable. Unlike Android, in iOS, the application code is compiled to machine code that can be analyzed using a disassembler.

The following are the major components of an iOS application package:

- `Info.plist`: Similar to the Android manifest file in an APK, this information property list file contains key-value pairs that specify essential runtime-configuration information for the application. The iOS operating system relies on the presence of this file to identify relevant information about the application and related files.

- **Executable**: The file that runs on the device, containing the application's main entry point and code that was statically linked to the application target.

- **Resource files**: Files that are required by the executable file, and are required for the application to properly run. This may contain images, nib files, string files, and configuration files.

The following diagram illustrates the iOS architecture overview:

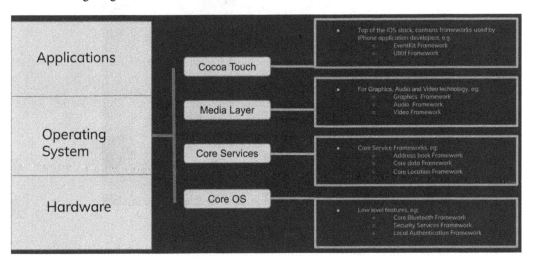

Figure 1.13 – iOS architecture

Let's see how to create a simple *hello world* application for iOS and then unzip it and look at its components:

1.  iOS apps are developed using Xcode. Download the latest version of Xcode from the App Store on Mac.

Figure 1.14 – Creating an Xcode project

2.  On the next screen, choose the default **App** template for your new project:

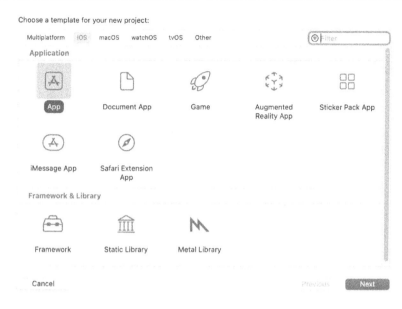

Figure 1.15 – Selecting the project template

3.  On the next screen, provide a product name (any name you like), select a team, and provide an organization identifier. To create and export an IPA from Xcode, you need to have an Apple Developer license:

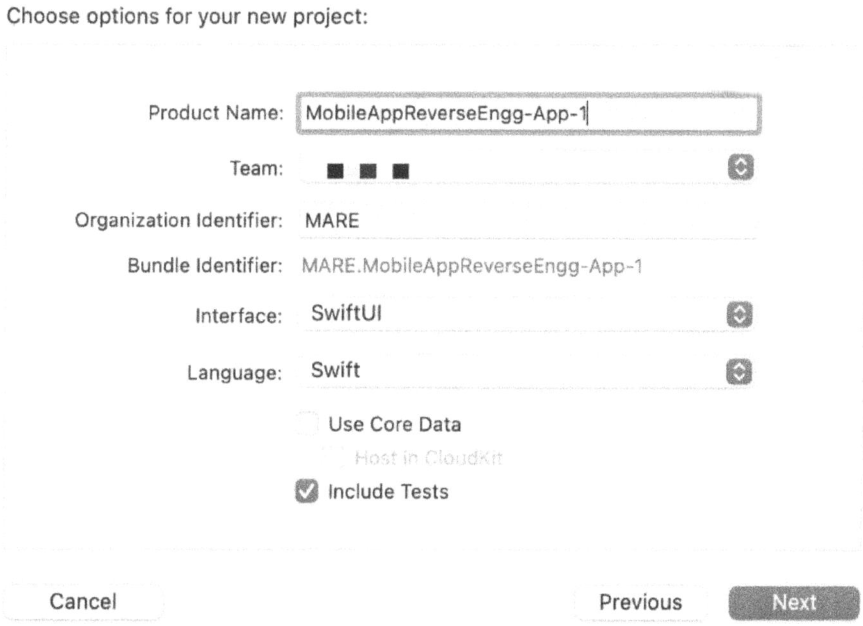

Choose options for your new project:

Product Name:  MobileAppReverseEngg-App-1

Team:  ■ ■ ■

Organization Identifier:  MARE

Bundle Identifier:  MARE.MobileAppReverseEngg-App-1

Interface:  SwiftUI

Language:  Swift

Use Core Data
Host in CloudKit
Include Tests

Cancel                    Previous    Next

Figure 1.16 – Providing project details

4.  Select a location to save the project on your computer.

    Xcode will now create a simple *hello world* application and you will see the following default code in the Xcode window:

Figure 1.17 – Project details

5.  Now you can try and run this app on one of the built-in iOS simulators. To do so, select one of the available simulators (just click on the name of simulator from top bar, and a list will open) as shown in the following screenshot:

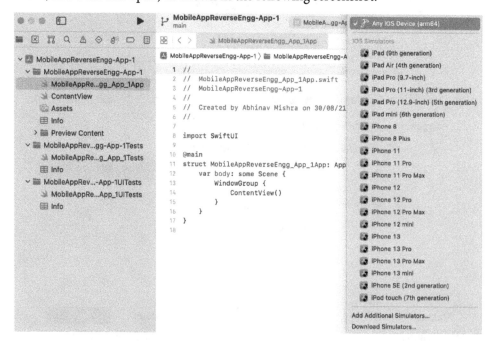

Figure 1.18 – Selecting a simulator

The app should run on the selected simulator:

Figure 1.19 – App running on the simulator

6.  Now, let's export the IPA from this Xcode project. To do so, select the **Any iOS Device (arm64)** option from the simulator options.

7.    Then, go to **Product | Archive** and select the **Distribute App** option:

Figure 1.20 – Exporting the application package

8.    On the next screen, select **Development** and leave the options on the subsequent screens at their defaults.

9.    Finally, you will be able to export the IPA together with some other compiled project files:

Review MobileAppReverseEngg-App-1.ipa content:

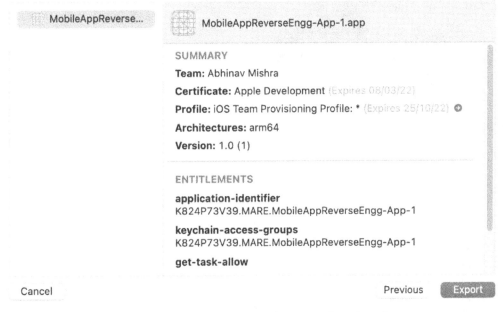

Figure 1.21 – Exporting the application package (cont.)

10. Once the IPA is exported, simply change the extension of the file to `.zip`:

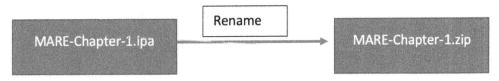

Figure 1.22 – Diagram explaining the application (IPA) extraction process via renaming

11. Use any tool to unzip the file and extract its contents:

```
# unzip MARE-Chapter-1.zip
```

The following screenshot shows the results for reference:

```
                 ~/Downloads     unzip MobileAppReverseEngg-App-1.ipa                            21:29:08
Archive:  MobileAppReverseEngg-App-1.ipa
  creating: Payload/
  creating: Payload/MobileAppReverseEngg-App-1.app/
  creating: Payload/MobileAppReverseEngg-App-1.app/_CodeSignature/
 inflating: Payload/MobileAppReverseEngg-App-1.app/_CodeSignature/CodeResources
 inflating: Payload/MobileAppReverseEngg-App-1.app/MobileAppReverseEngg-App-1
 inflating: Payload/MobileAppReverseEngg-App-1.app/embedded.mobileprovision
 inflating: Payload/MobileAppReverseEngg-App-1.app/Info.plist
 inflating: Payload/MobileAppReverseEngg-App-1.app/PkgInfo
```

Figure 1.23 – Extracting the content of the IPA after renaming it to ZIP

12. Go into the `Payload` directory and then inside the `MobileAppReverseEngg-App-1.app` file:

```
# cd Payload
# cd MobileAppReverseEngg-App-1.app
```

13. Let's analyze the components inside the IPA and compare it with the list here (*iOS application fundamentals*):

```
0ctac0der@Mac  ~/Downloads/Payload/MobileAppReverseEngg-App-1.app  ls -l          21:30:29
total 240
-rw-r--r--@ 1 0ctac0der  staff   2572 Oct 25 21:23 Info.plist
-rwxr-xr-x@ 1 0ctac0der  staff  91280 Oct 25 21:23 MobileAppReverseEngg-App-1
-rw-r--r--@ 1 0ctac0der  staff      8 Oct 25 21:23 PkgInfo
drwxr-xr-x@ 3 0ctac0der  staff     96 Oct 25 21:23
-rw-r--r--@ 1 0ctac0der  staff  17970 Oct 25 21:23 embedded.mobileprovision
```

Figure 1.24 – Extracted content of the IPA

The following diagram illustrates the process of reverse engineering an iOS application:

Figure 1.25 – Overview of the reverse engineering process of an IPA

Have a look at *Figure 1.3* to understand how a disassembled binary looks in Hopper disassembler.

## Summary

The concepts and processes of reverse engineering are very interesting. Through this chapter, you have learned the fundamentals of reverse engineering both Android and iOS applications. The concepts explored will help your understanding in the later chapters of this book as we begin to look at reverse engineering in depth.

In the next chapter, we will learn more about the modern tools used to reverse engineer iOS and Android apps.

# 2
# Setting Up a Mobile App Reverse Engineering Environment Using Modern Tools

As you already understand the fundamentals of reverse engineering, let's start exploring some modern reverse engineering tools that can be used for mobile applications.

In order to reverse engineer mobile apps, we need specialized tools and utilities. Some of those tools are paid and some are open source. We will try to use the open source tools and utilities as much as possible, but will also provide you with a commercial alternative, wherever applicable.

Before we start setting up the tools in our newly created Ubuntu virtual machine, it is important to understand that some of the tools work for both iOS and Android apps, while some of them only work on one. For each tool explained, you will find a *Use case example* section. This section will provide information about what platform this tool will work on, that is, iOS/Android or both.

At the end of the chapter, you will also be provided with details about a customized virtual machine platform for penetration testing and the reverse engineering of mobile applications, Mobexler.

This chapter talks about some of the fun little utilities, as well as commercial tools, that can be used for reverse engineering.

In this chapter, we will cover the following topics:

- Tools for the reverse engineering of mobile (Android and iOS) applications
- Setting up an environment for reverse engineering
- Installing and setting up the tools for reverse engineering
- Setting up Mobexler (a mobile application penetration testing platform)

# Technical requirements

Download and set up a virtual machine (at the time of writing this book, Ubuntu 20.04.3 LTS is the latest version, and we will be using that) using any virtualization software, such as VirtualBox or VMware. You can download Ubuntu 20.04.3 LTS (Ubuntu desktop) at `https://ubuntu.com/download/desktop`.

For virtualization, you can use the open source VirtualBox (`https://www.virtualbox.org/`) or the commercially available version (as well as the free version) of VMware Workstation Player (`https://www.vmware.com/in/products/workstation-player.html`) for Windows, or VMware Player Fusion (`https://www.vmware.com/in/products/fusion.html`) for Mac.

The steps to download and set up a virtual machine are not covered, as it is a straightforward and easy-to-do task. Following any good article/blog post on how to set up an Ubuntu virtual machine should provide all necessary information. Here is a post from the VirtualBox official website about creating a Windows-based virtual machine: `https://www.virtualbox.org/manual/ch01.html#gui-createvm`.

# Tools for the reverse engineering of mobile applications

We learned in the last chapter that Android apps, as well as iOS apps, come in a specific format (APK or IPA), which is nothing but a compressed (.zip) version of all the application files and most importantly the compiled binary file.

When we start with the reverse engineering of mobile apps, the primary goal is to understand the internals of the application, including its features and implemented security controls, and reconstruct as much original code as possible. To do this in a mobile application, the first step is to decompress or, more specifically, decompile the application package itself.

When you start, the first step is to get the application package (APK or IPA) and decompress it. To do that, you need a simple utility that decompresses a compressed file (.zip). Some such utilities come preinstalled with most Linux operating systems.

Just start your newly created Ubuntu virtual machine and start Terminal.

To use the unzip utility, type the following in Terminal:

```
# unzip --help
```

The preceding command should result in running the unzip utility and showing you all the different options available.

Now, let's try and unpack an APK file. Simply take the APK file we created in the previous chapter and try to unpack it using the unzip utility. Make sure the APK file is saved in the Desktop folder, or create a new folder and save the file in that folder. Once the file is saved, open Terminal in that folder by right-clicking somewhere inside it and selecting **Open in Terminal**.

1. Rename the APK file to ZIP:

   ```
   #mv app-debug.apk app-debug.zip
   ```

2. Extract the files from the ZIP file using the unzip utility:

   ```
   #unzip app-debug.zip
   ```

The extracted files are as follows:

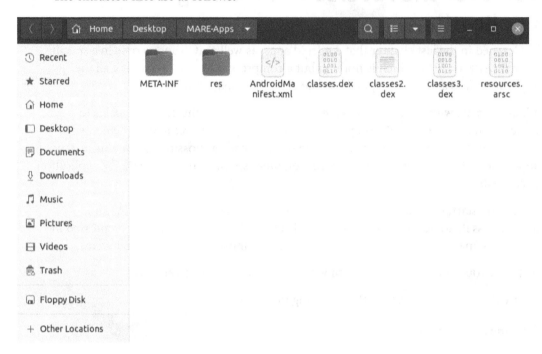

Figure 2.1 – Running the unzip utility to unzip an APK file

Even though you can unzip the APK/IPA and explore the files inside it, for Android, this is not the correct way to start with reverse engineering. For example, once extracted, if you try and view the `AndroidManifest.xml` file, it will not be readable. This is because at this point, you have simply inflated compiled sources, and editing or viewing a compiled file is not that easy.

This is where the use of the `apktool` tool comes into play. To start with the reverse engineering of an Android application, the correct approach is by using `apktool`, which can properly decode APK resources to almost the original form and rebuild them after making some modifications. The decode option in `apktool` will convert important files, such as config and resources, to XMLs.

As we just saw, the reverse engineering of iOS and Android apps requires very specific modern tools. So, let's look at some of those tools and set them up in a virtual machine environment. After the installation and setup of all these tools, we will have a fresh, ready-to-use environment for the reverse engineering of mobile apps (both Android and iOS apps).

# apktool

Tool: `apktool`

Website: `https://ibotpeaches.github.io/Apktool/install/`

About: A tool for reverse engineering Android apps

Used for: Android apps

Here are the instructions to install `apktool` (in Ubuntu):

1. Download the Linux wrapper script (right-click and select **Save Link As apktool**) from this link: `https://raw.githubusercontent.com/iBotPeaches/Apktool/master/scripts/linux/apktool`:

   ```
   #wget https://raw.githubusercontent.com/iBotPeaches/
   Apktool/master/scripts/linux/apktool
   ```

2. Download `apktool-2` (find the current version, 2.6.0): `https://bitbucket.org/iBotPeaches/apktool/downloads/`.

3. Rename the downloaded JAR to `apktool.jar`.

4. Move both files (`apktool.jar` and `apktool`) to `/usr/local/bin` (root needed).

5. Make sure both files are executable (`chmod +x`).

6. Try running `apktool` via the CLI:

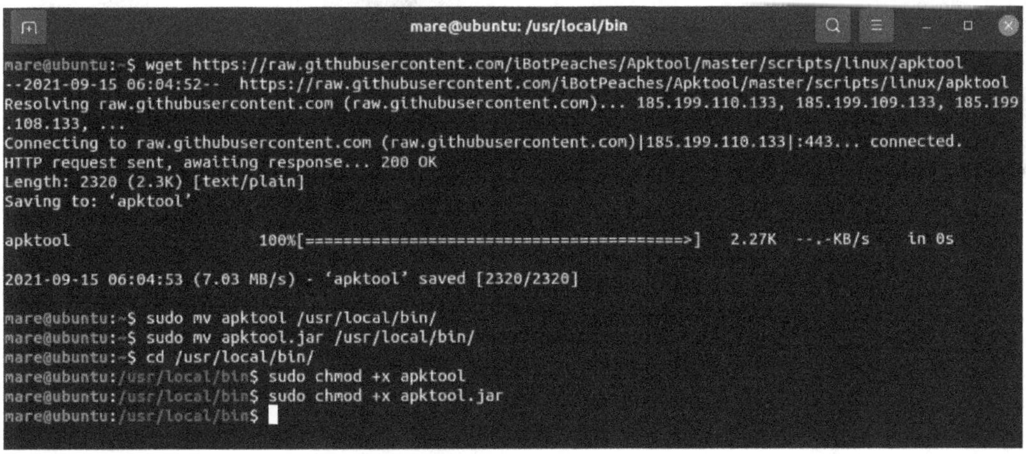

Figure 2.2 – Setting up apktool

If you see an error related to Java, it might be possible that Java is not installed. In that case, please install Java with the following command:

```
#sudo apt install default-jdk
#sudo apt install default-jre
```

## Use case example

apktool is probably the first tool you will use when starting to reverse engineer an Android application package. apktool is used to first decode the original Android package and extracts all the files from inside it together with the dex files.

Let's have a look at how it can be used to decode an APK and get all its components:

1.  Once installed, we will simply start Terminal and type the following command:

    ```
    #apktool d <APK Path>
    ```

    Here, d stands for *decode*.

    Here is the snippet for reference:

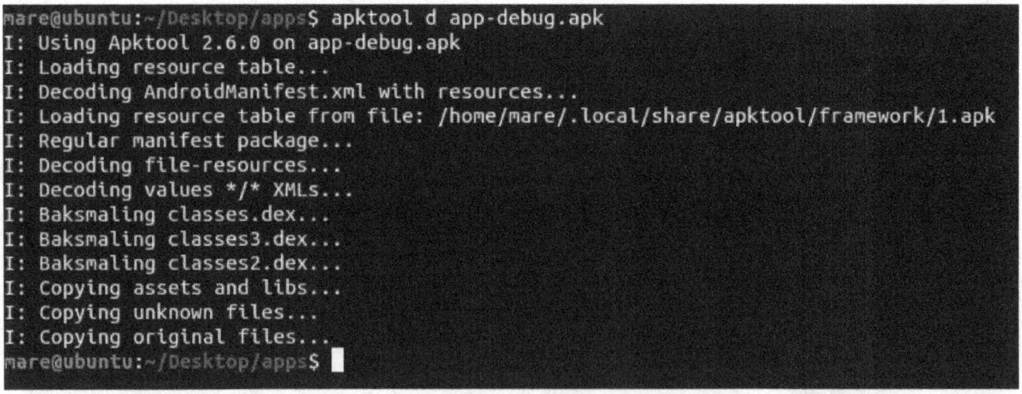

```
mare@ubuntu:~/Desktop/apps$ apktool d app-debug.apk
I: Using Apktool 2.6.0 on app-debug.apk
I: Loading resource table...
I: Decoding AndroidManifest.xml with resources...
I: Loading resource table from file: /home/mare/.local/share/apktool/framework/1.apk
I: Regular manifest package...
I: Decoding file-resources...
I: Decoding values */* XMLs...
I: Baksmaling classes.dex...
I: Baksmaling classes3.dex...
I: Baksmaling classes2.dex...
I: Copying assets and libs...
I: Copying unknown files...
I: Copying original files...
mare@ubuntu:~/Desktop/apps$
```

Figure 2.3 – Using apktool

2.  When the decoding completes, all the components of the APK will be saved inside a folder with the same name as the package:

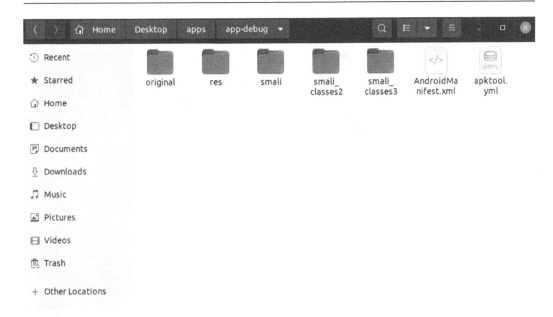

Figure 2.4 – Decoded application components

# JADX – Dex-to-Java decompiler

Tool: JADX – Dex-to-Java decompiler

Website: https://github.com/skylot/jadx

About: Command line and GUI tools for producing Java source code from Android dex and APK files

Used for: Android apps

Follow these instructions to install it (in Ubuntu):

1.  Download the latest available version of the tool from https://github.com/skylot/jadx/releases/.

2.  The downloaded file will be a ZIP; extract it to a folder.

3.  Navigate to the folder where you have extracted the file and go inside the bin folder.

4.  To use the GUI version of the application, run the following:

```
#./jadx-gui
```

The result is as follows:

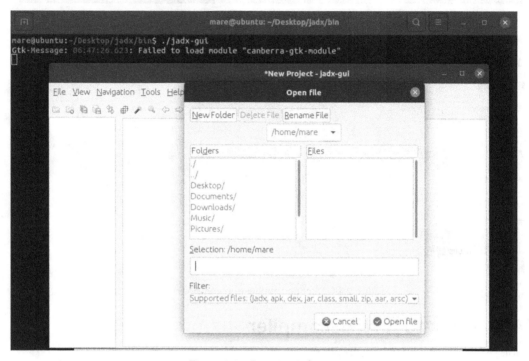

Figure 2.5 – Running jadx-gui

## Use case example

JADX is a great tool, as it takes an Android package (APK) as the input and then provides the Java source code as the output. It takes care of decoding the APK and then converting the dex files to JAR files, which are then interpreted in the reader.

Let's load the APK we created directly into the jadx-gui application and look at the output:

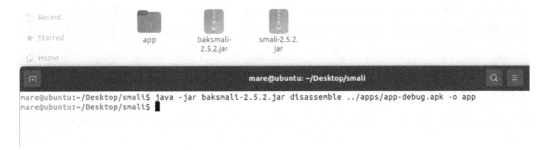

Figure 2.6 – JADX showing the Java source code from the APK

# smali/baksmali

Tool: smali/baksmali

Website: `https://github.com/JesusFreke/smali`

About: An assembler/disassembler for the `dex` format used by Dalvik

Used for: Android apps

Follow these instructions to install it (in Ubuntu):

1. Download and save the latest stable version of the tool from `https://bitbucket.org/JesusFreke/smali/downloads/`.

2. Use Java to run either of the apps. For example, run `baksmali` as shown in the following figure:

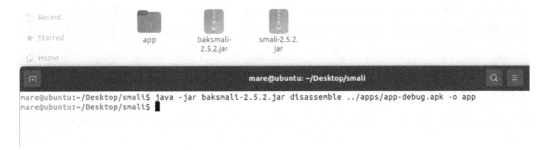

Figure 2.7 – Using baksmali to disassemble an APK

# strings

Tool: `strings`

Website: `https://www.gnu.org/software/binutils/`.

About: The Linux `strings` command is primarily used for finding string characters in files. It focuses on determining the contents of, and extracting text from, binary files (non-text files). Different operating systems might have different arguments.

Used for: Android and iOS apps

To install it (in Ubuntu), run the following command in Terminal:

```
#sudo apt-get install binutils
```

The screenshot for reference is as follows:

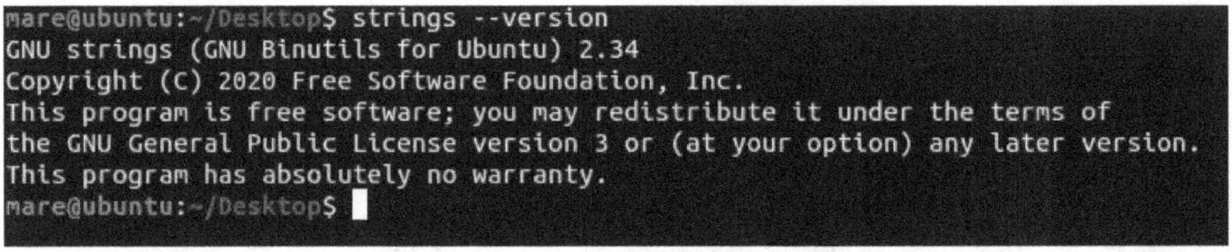

```
mare@ubuntu:~/Desktop$ strings --version
GNU strings (GNU Binutils for Ubuntu) 2.34
Copyright (C) 2020 Free Software Foundation, Inc.
This program is free software; you may redistribute it under the terms of
the GNU General Public License version 3 or (at your option) any later version.
This program has absolutely no warranty.
mare@ubuntu:~/Desktop$
```

Figure 2.8 – Running strings

## Use case example

The `strings` utility is very helpful when you are trying to find a static string in the application binary. Just pass the binary as input to the `strings` utility and it will extract all the strings from it. Looking at the extracted strings is very helpful, as it can provide the class names, method names, static text, and hardcoded information, for example.

Let's zip extract (change the extension to `.zip` and extract using any zip extractor tool) the APK file we have, and then run `strings` on the `classes.dex` file. This should extract all the strings inside the `dex` file.

Run the `$strings classes.dex` command and it will extract all the strings. You can also save all the extracted strings by using output redirection with `$strings classes.dex >> extracted_strings.txt`.

# Ghidra

Tool: Ghidra

Website: `https://github.com/NationalSecurityAgency/ghidra/releases`.

About: A **software reverse engineering** (**SRE**) suite of tools developed by the NSA's Research Directorate

Used for: Primarily iOS apps

Follow these instructions to install it (in Ubuntu):

1. Download the latest release from `https://github.com/NationalSecurityAgency/ghidra/releases`.

2. Extract the ZIP file to any folder.

3. Navigate to the folder where the ZIP file was extracted and run the following command in Terminal:

```
#./ghidraRun
```

The screenshot for reference is as follows:

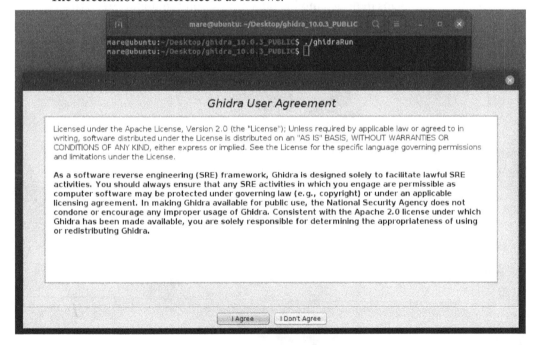

Figure 2.9 – Running Ghidra

4.  Once you start Ghidra, create a project and then choose one of the tools to run, as shown here:

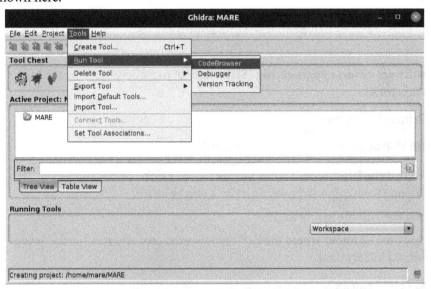

Figure 2.10 – Starting the CodeBrowser Ghidra tool

The CodeBrowser tool helps in disassembling binary files. You can import the `classes.dex` file (just drag and drop) and the tool will show the disassembled code.

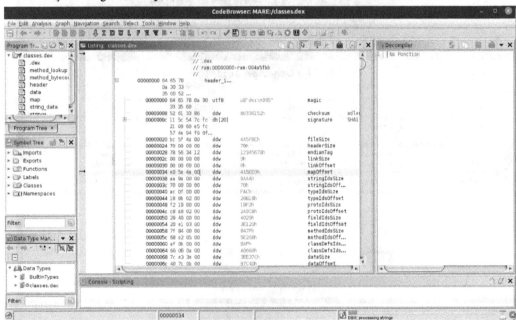

Figure 2.11– Starting the CodeBrowser Ghidra tool

# Radare

Tool: Radare

Website: https://rada.re/r/.

About: Disassembling (and assembling) many different architectures

Used for: Primarily iOS apps

Follow these instructions to install it (in Ubuntu):

1.  Run the following commands in Terminal:

    ```
    #sudo apt-get update
    #sudo apt-get install radare2
    ```

2.  Once installed, run the following command to confirm:

    ```
    #radare2 -version
    ```

    The screenshot for reference is as follows:

    ```
    mare@ubuntu:~/Desktop/ghidra_10.0.3_PUBLIC$ radare2 -version
    radare2 4.2.1 0 @ linux-x86-64 git.4.2.1
    commit: unknown build:
    mare@ubuntu:~/Desktop/ghidra_10.0.3_PUBLIC$
    ```

Figure 2.12 – Running radare2

So far, we have completed the installation and setup of all the main tools required for the reverse engineering of mobile apps, although some more tools might be needed in the chapters ahead. We will install those utilities/tools when required.

Installing and updating the various tools and utilities that are required for reverse engineering can be a big task. Penetration testers and reverse engineers need a proper setup for their work every day. What if there was a customized, well-set-up, and prebuilt platform for the security testing and reverse engineering of mobile apps? Well, here comes **Mobexler**, a mobile application penetration testing platform.

Mobexler comes preinstalled with all the necessary tools and utilities that are required by penetration testers and reverse engineers. Mobexler is specifically made to help in the security testing of Android apps as well as iOS apps.

However, when starting with any new topic, it is always best to spend some extra time and set up everything manually. We will be using the Ubuntu virtual machine environment for the rest of the book, but once you are comfortable with all the tools and understand their usage, feel free to use a pre-set-up environment, such as Mobexler.

# Mobexler virtual machine

About: A mobile application penetration testing platform that comes preinstalled with all the tools required for the reverse engineering and penetration testing of Android and iOS applications

Website: `https://mobexler.com/`

Used for: Android and iOS

Follow these instructions to set it up:

1.  Download the latest image (`ova`) file from the Mobexler website: `https://mobexler.com/download.htm`.

2.  Import the downloaded virtual machine image in virtualization software, such as VirtualBox or VMware.

3.  You can also follow the step-by-step guide on the Mobexler website: `https://mobexler.com/setup.htm`.

4.  Once imported as a virtual machine, log in to the virtual machine using the password mentioned in the virtual machine description.

To learn more about the tools installed inside Mobexler, you can visit the Mobexler page at `https://mobexler.com/tools.htm`.

## Use case example

As mentioned, Mobexler comes preinstalled with all the main tools you need. So, it can be used for almost all kinds of mobile apps, such as Java, Kotlin, Swift, and Objective C. For example, let's take the same APK we created in the last chapter and try to use different tools from within Mobexler.

In the following image, we are going to run `apktool`, which is preinstalled in Mobexler, and decode an APK file. The APK file used is the same sample APK created in the previous section of this book. You can use the same command to decompile any other APK as well.

The command is as follows:

```
#apktool d [apk_nam/path]
```

Here, d stands for *decode*.

The screenshot for reference is as follows:

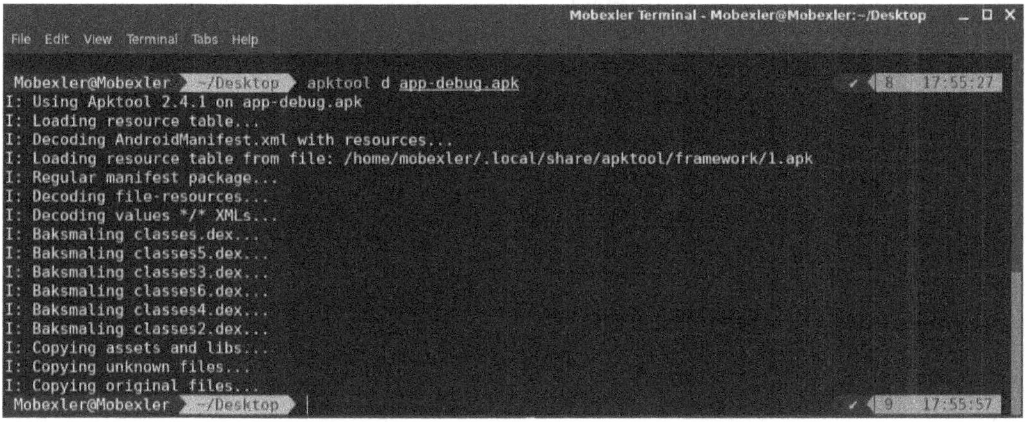

```
Mobexler Terminal - Mobexler@Mobexler:~/Desktop    _ □ X
File  Edit  View  Terminal  Tabs  Help

Mobexler@Mobexler  ~/Desktop  apktool d app-debug.apk              8    17:55:27
I: Using Apktool 2.4.1 on app-debug.apk
I: Loading resource table...
I: Decoding AndroidManifest.xml with resources...
I: Loading resource table from file: /home/mobexler/.local/share/apktool/framework/1.apk
I: Regular manifest package...
I: Decoding file-resources...
I: Decoding values */* XMLs...
I: Baksmaling classes.dex...
I: Baksmaling classes5.dex...
I: Baksmaling classes3.dex...
I: Baksmaling classes6.dex...
I: Baksmaling classes4.dex...
I: Baksmaling classes2.dex...
I: Copying assets and libs...
I: Copying unknown files...
I: Copying original files...
Mobexler@Mobexler  ~/Desktop  |                                    9    17:55:57
```

Figure 2.13 – Using apktool in Mobexler to decode an APK

Similarly, you can use other more advanced tools, such as radare2, right from the Mobexler Terminal:

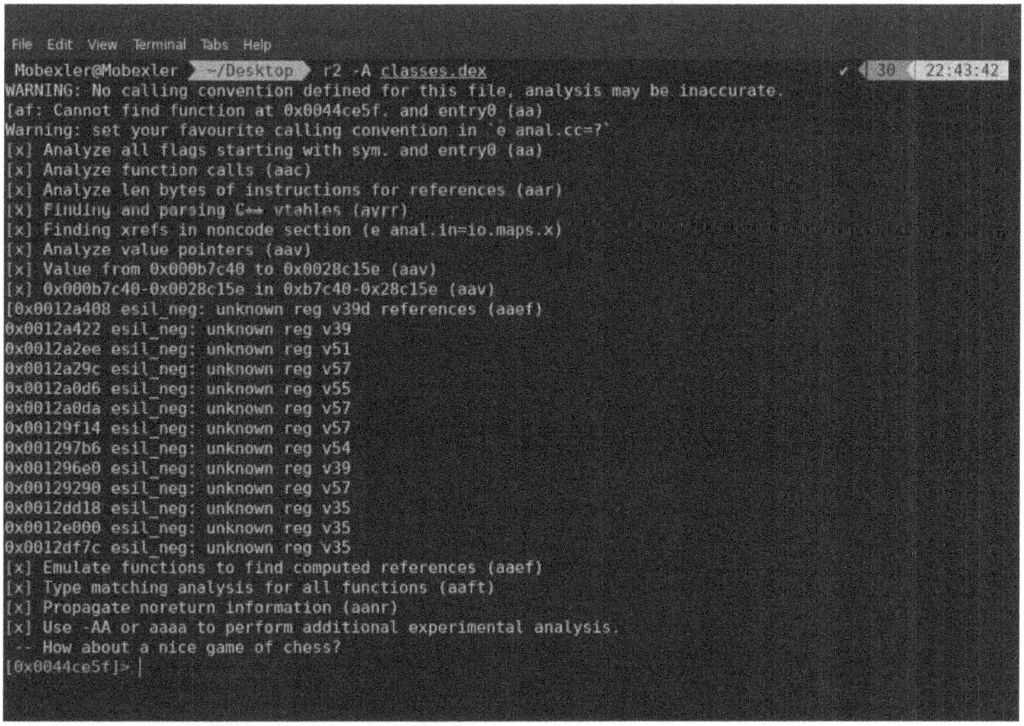

```
File  Edit  View  Terminal  Tabs  Help
Mobexler@Mobexler  ~/Desktop  r2 -A classes.dex                    30    22:43:42
WARNING: No calling convention defined for this file, analysis may be inaccurate.
[af: Cannot find function at 0x0044ce5f. and entry0 (aa)
Warning: set your favourite calling convention in `e anal.cc=?`
[x] Analyze all flags starting with sym. and entry0 (aa)
[x] Analyze function calls (aac)
[x] Analyze len bytes of instructions for references (aar)
[x] Finding and parsing C++ vtables (avrr)
[x] Finding xrefs in noncode section (e anal.in=io.maps.x)
[x] Analyze value pointers (aav)
[x] Value from 0x000b7c40 to 0x0028c15e (aav)
[x] 0x000b7c40-0x0028c15e in 0xb7c40-0x28c15e (aav)
[0x0012a408 esil_neg: unknown reg v39d references (aaef)
0x0012a422 esil_neg: unknown reg v39
0x0012a2ee esil_neg: unknown reg v51
0x0012a29c esil_neg: unknown reg v57
0x0012a0d6 esil_neg: unknown reg v55
0x0012a0da esil_neg: unknown reg v57
0x00129f14 esil_neg: unknown reg v57
0x001297b6 esil_neg: unknown reg v54
0x001296e0 esil_neg: unknown reg v39
0x00129290 esil_neg: unknown reg v57
0x0012dd18 esil_neg: unknown reg v35
0x0012e000 esil_neg: unknown reg v35
0x0012df7c esil_neg: unknown reg v35
[x] Emulate functions to find computed references (aaef)
[x] Type matching analysis for all functions (aaft)
[x] Propagate noreturn information (aanr)
[x] Use -AA or aaaa to perform additional experimental analysis.
-- How about a nice game of chess?
[0x0044ce5f]> |
```

Figure 2.14 – Using apktool in Mobexler

Now that we have got the environment fully set up for the reverse engineering of mobile apps, we will start working on some exercises in which we will reverse engineer real-world mobile apps. In order to help you understand the concepts better, we will be providing the source code (as well as the application package) of a real-world application. This application will be available in multiple formats, that is, Java and Kotlin for Android applications, and Swift and Objective C for iOS apps.

# Summary

In this chapter, we looked at some of the awesome utilities and tools that can be used for the purpose of reverse engineering Android and iOS applications. Knowing how to use these tools and when to use them helps in properly reverse engineering apps. As you have a reverse engineering environment ready with all the tools, either using an Ubuntu virtual machine or using Mobexler, we can now proceed to actually use these tools and reverse engineer a real Android or iOS application.

In the next chapter, we will start with a more in-depth analysis of a Java-based Android application, where we will try to reverse engineer the application, look inside the source code, and understand its different features.

For the purpose of this book, we have created an Android, as well as an iOS, version of an app called **SecureStorage**. In the next chapter, let's look at what the SecureStorage app is about and how we can reverse engineer it.

# Section 2: Mobile Application Reverse Engineering Methodology and Approach

This section covers Android and iOS apps one by one in separate chapters to look into how those apps are developed, exploring the internals of the apps and the process to reverse engineer them. This section covers Android apps, iOS apps built using Objective-C, and iOS apps built using Swift, ranging from unzipping application packages to extracting content, as well as binaries. We will look into the process of how to reverse engineer an application's binary and understand the workings of an application(s).

This part of the book comprises the following chapters:

- *Chapter 3, Reverse Engineering an Android Application*
- *Chapter 4, Reverse Engineering an iOS Application*
- *Chapter 5, Reverse Engineering an iOS Application (Developed Using Swift)*

# 3
# Reverse Engineering an Android Application

In the last two chapters, you learned about the basics of reverse engineering and looked into some of the tools used and their installation. You should now be able to create an Ubuntu-based virtual machine environment (or have already done so). Then, you learned how to install and run the reverse engineering tools listed in *Chapter 2, Setting Up a Mobile App Reverse Engineering Environment Using Modern Tools* (only some of the basic operation of the tools was covered, not all the features).

In this chapter, we will be covering the following:

- Android application development
- The reverse engineering of Android applications
- Extracting Java source code
- Converting .dex files to smali
- Reverse engineering and penetration testing
- Code obfuscation in Android apps

# Technical requirements

This chapter has the following technical requirements:

- An Ubuntu virtual machine with the tools listed in *Chapter 2, Setting Up a Mobile App Reverse Engineering Environment Using Modern Tools* (only Android-specific tools).

- Download the SecureStorage Android application source code from this link: `https://github.com/0ctac0der/SecureStorage-AndroidApp`

# Android application development

Before we get into reverse engineering, it is important to understand how the forward-engineering process happens and how an application is developed. In order to create an application, the developer chooses a programming language according to the operating system on which the application is supposed to run. For example, in the case of Android applications, generally, a developer may choose to develop their application using Java, Kotlin, or C++ (using other languages is also possible). Kotlin is the official language for Android applications. Android Studio is the official development environment for building Android apps. We already used Android Studio in the *Android application fundamentals* section of *Chapter 1, Basics of Reverse Engineering – Understanding the Structure of Mobile Apps*. Android Studio contains a comprehensive set of development tools, such as **Android Debug Bridge** (**ADB**), **fastboot**, and the **Native Development Kit** (**NDK**). The Android **software development kit** (**SDK**) is fully integrated with Android Studio and can be easily installed using the SDK Manager. However, we can also use the Android SDK independently without Android Studio. Refer to the following screenshot, which shows the Android SDK as a part of Android Studio:

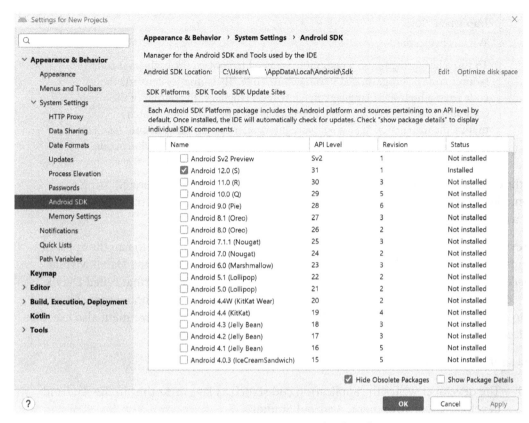

Figure 3.1 – SDK Manager in Android Studio

The developer writes the required code for all the functionalities of the application, its user interface, and the logic for data processing. Together with the code, all the required resource files, configuration files, and images are also added as a part of the project in Android Studio. Then, this code (together with the resources) is compiled using the SDK.

Let's say that the code is written in Java. It will be compiled by the Java compiler (javac) into Java bytecode files (class files). Then, these Java bytecode files are sent to a **Java Virtual Machine (JVM)**, which converts them into machine code using a **just-in-time (JIT)** compiler and runs them on the device.

> **Important Note**
>
> What is bytecode?
>
> To run a Java program on a computer, we need to convert the high-level code (source code or program code) to machine code. The compiler converts the code from a high-level language to a low-level language; the output of the compilation process is bytecode. Bytecode is the low-level code, which is mainly the instruction set for software interpreters or virtual machines (such as JVM). Finally, the bytecode is translated by an interpreter into machine code.

If the code is written in Kotlin, Java-compatible bytecode can be generated using a compiler, such as `kotlinc`. Bytecode is nothing but a form of an instruction set executed by a software interpreter.

Previously, Android versions 1.0 to 4.4 had a Dalvik virtual machine to run the apps. In Android 4.4, Google added **Android Runtime** (**ART**) together with the Dalvik virtual machine. The Dalvik virtual machine runs an optimized bytecode format called Dalvik bytecodes. In the compilation process, the `.class` files and `.jar` libraries are converted into `classes.dex` files containing Dalvik bytecode. The ART environment also executes `.dex` files.

Let's summarize this as follows:

1. The developer writes the application code (such as Java or Kotlin) in the Android development environment, Android Studio.

2. The written code is compiled to Dalvik executable (`.dex`) files using DEX compilers, which will run in the ART environment (as well as the Dalvik virtual machine).

3. The compilers in Android Studio will also compile the other resource files, JAR libraries, and other libraries (if any).

4. The compiled `.dex` files and the compiled resources will be packed together to create the **Android Package** (**APK**).

> **Important Note**
>
> For an Android application developed using Java, the code is compiled to DEX bytecode. The reverse engineering process works in the opposite direction – extract the `.dex` files from the APK and convert them to Java code.

The final application package (meaning the APK) will contain the following important files and folders, together with other entities:

- Resource files in the `res` folder
- `AndroidManifest.xml`
- A `resources.arsc` file, which contains all meta-information
- `.dex` files

In *Chapter 2, Setting Up a Mobile App Reverse Engineering Environment Using Modern Tools*, when we simply extracted all the content of an APK file by changing its extension to `zip`, we found the files listed in the preceding list. Refer to *Chapter 2, Figure 2.1 – Running the unzip utility to unzip an APK file.*

The Android operating system requires that all apps be digitally signed with a certificate before they can be installed. During the installation process, the Android operating system uses the Package Manager to verify that the APK has been properly signed with the certificate included in that APK. Developers can use self-signed certificates for signing the applications that they develop. A developer generates a certificate and uses it to sign the application before a release build is generated. The certificate file is also a part of the APK.

# The reverse engineering of Android applications

Let's now look at the ways to reverse engineer Android applications and study the bits that can be extracted from a compiled application package.

In this section, we will be using a specially built application called **SecureStorage**. You can download the different builds of the application from the following GitHub links:

- The debug build of SecureStorage (`https://github.com/0ctac0der/SecureStorage-AndroidApp/releases/download/0.1/app-debug.apk`)
- The release build of SecureStorage (`https://github.com/0ctac0der/SecureStorage-AndroidApp/releases/download/1.0/app-release.apk`)

SecureStorage is a simple Android application that can be used to store credit card information on the device. A user will have to sign in with the correct password to be able to access the stored information.

You can install the downloaded application on the Android virtual device to see how it works. Some of the screens look as follows:

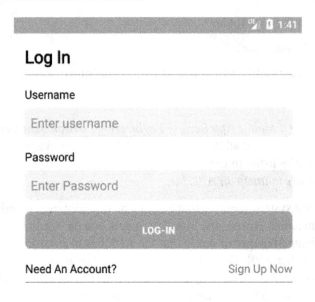

Figure 3.2 – Home screen of the SecureStorage app

If the user does not have an account in the app, they can create an account from the **Join Today** screen, which looks as follows:

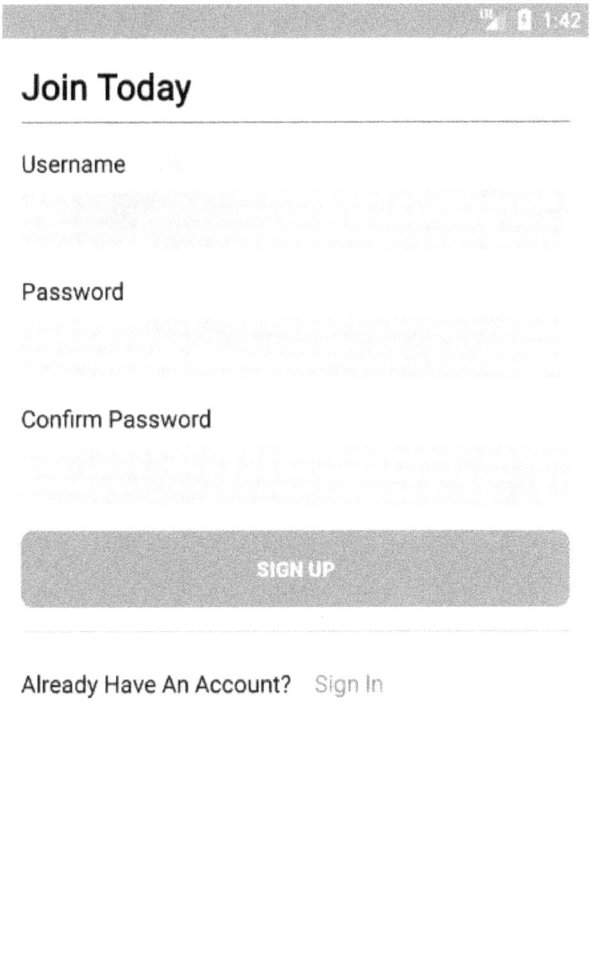

Figure 3.3 – Join Today screen of the SecureStorage app

Once logged in, users get multiple options to save a credit card, modify an already stored card, and view previously added cards. The following screen shows all those options:

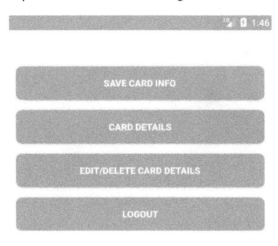

Figure 3.4 – Saved details screen in the SecureStorage app

**Important Points on the SecureStorage App**

The application works fully on the client side, which means there is no backend or server side for this application.

As the name suggests, SecureStorage tries to securely store data in the user's device storage.

So far, we have looked into details of how Android apps are developed and run on a device, and their internal components, and have also installed an app (SecureStorage). Now, let's move forward to start reverse engineering the app and look at what's hidden inside.

# Extracting the Java source code

The first objective of reverse engineering is to get the original source code with maximum accuracy. As we have the application package downloaded on our Ubuntu virtual machine, let's use the **JADX** tool to get the Java code.

However, it might also be a good idea to simply unzip the APK and extract its contents to see what's inside:

Figure 3.5 – Extracted contents of the APK

In order to use the JADX tool, open the directory where you extracted the JADX .zip file (as explained in *Chapter 2, Setting Up a Mobile App Reverse Engineering Environment Using Modern Tools*). Once in the directory, right-click to select the **Open in Terminal** option. In the opened Terminal window, type the following command to run JADX:

```
# cd bin/
# ./jadx-gui
```

In the JADX window, open the APK file you just downloaded. Refer to *Chapter 2, Setting Up a Mobile App Reverse Engineering Environment Using Modern Tools*, to see how to do this:

```
- app-debug.apk
  - Source code          AndroidManifest.xml ×
    - android.support.v4      <?xml version="1.0" encoding="utf-8"?>
    - androidx              2<manifest xmlns:android="http://schemas.android.com/apk/res/android" android:versionCode="1"
    - com                   7    <uses-sdk android:minSdkVersion="21" android:targetSdkVersion="30"/>
  - Resources             11   <application android:theme="@style/Theme.SecureStorage" android:label="@string/app_name"
    - META-INF            21        <activity android:name="com.example.securestorage.activity.LoginActivity">
    - res                 22            <intent-filter>
    - AndroidManifest.xml 23                <action android:name="android.intent.action.MAIN"/>
    classes.dex          25                <category android:name="android.intent.category.LAUNCHER"/>
    classes2.dex         22            </intent-filter>
    classes3.dex         21        </activity>
    classes4.dex         28        <activity android:name="com.example.securestorage.activity.HomeActivity"/>
    classes5.dex         29        <activity android:name="com.example.securestorage.activity.SignupActivity"/>
    classes6.dex         30        <activity android:name="com.example.securestorage.activity.SaveCardInfoActivity"/>
    resources.arsc       31        <activity android:name="com.example.securestorage.activity.CardDetailsActivity"/>
  APK signature          11   </application>
                          2</manifest>
```

Figure 3.6 – Decompiled Java source code of the application

In the majority of applications, it is possible to reverse engineer the application package to the decompiled Java source code. However, in some cases, it is possible that the decompiled Java code does not look very clear, or multiple parts of the Java code are not readable at all in the Java decompilers.

> **Important Note**
> The decompiled Java code from JADX is generally not recompilable.

When code is highly obfuscated, or the Java code is difficult to obtain, we can convert the .dex files (Dalvik bytecode) to smali. The .dex files contain the binary Dalvik bytecode, which is not at all easy to read or modify, so it is useful to convert that bytecode into a more human-readable format, smali. So, it is an approach to understand the code through the decompiled code from JADX and use the smali version to edit any part of it and then recompile it. A pair of tools called **smali** and **baksmali**, which are technically an assembler and a disassembler, can be used to convert the smali code to .dex format, and .dex to smali, respectively.

## Converting DEX files to smali

Let's try to convert the same APK to smali files, using another tool we used in *Chapter 2, Setting Up a Mobile App Reverse Engineering Environment Using Modern Tools.*

In order to decompile the APK, run the following command:

```
# apktool d app-debug.apk
```

apktool uses smali/baksmali internally, while decompiling an APK file. The following figure shows that apktool is decoding the app-debug.apk file provided:

```
mare@ubuntu:~/Desktop/app$ apktool d app-debug.apk
I: Using Apktool 2.6.0 on app-debug.apk
I: Loading resource table...
I: Decoding AndroidManifest.xml with resources...
I: Loading resource table from file: /home/mare/.local/share/apktool/framework/1.apk
I: Regular manifest package...
I: Decoding file-resources...
I: Decoding values */* XMLs...
I: Baksmaling classes.dex...
I: Baksmaling classes3.dex...
I: Baksmaling classes6.dex...
I: Baksmaling classes4.dex...
I: Baksmaling classes5.dex...
I: Baksmaling classes2.dex...
I: Copying assets and libs...
I: Copying unknown files...
I: Copying original files...
mare@ubuntu:~/Desktop/app$ ▉
```

Figure 3.7 – Using apktool to decompile the application

Once the APK has been decompiled, navigate to the folder created (in this case app-debug), and you will find several subfolders inside it with the name smali*. These folders contain the converted smali files from the .dex files in the APK:

```
mare@ubuntu:~/Desktop/app/app-debug$ ls -l
total 40
-rw-rw-r--    1 mare mare 1357 Dec  6 01:05 AndroidManifest.xml
-rw-rw-r--    1 mare mare 2564 Dec  6 01:06 apktool.yml
drwxrwxr-x    3 mare mare 4096 Dec  6 01:06 original
drwxrwxr-x  150 mare mare 4096 Dec  6 01:05 res
drwxrwxr-x    5 mare mare 4096 Dec  6 01:06 smali
drwxrwxr-x    4 mare mare 4096 Dec  6 01:06 smali_classes2
drwxrwxr-x    3 mare mare 4096 Dec  6 01:06 smali_classes3
drwxrwxr-x    3 mare mare 4096 Dec  6 01:06 smali_classes4
drwxrwxr-x    3 mare mare 4096 Dec  6 01:06 smali_classes5
drwxrwxr-x    3 mare mare 4096 Dec  6 01:06 smali_classes6
mare@ubuntu:~/Desktop/app/app-debug$
```

Figure 3.8 – Decompiled content from apktool

Opening any of the smali files will show the respective version of the code. Let's look at the content of the smali files for the classes5.dex file. To do so, we will need to navigate to the smali_classes5/com/example/securestorage/adapter directory:

```
# cd smali_classes5/com/example/securestorage/adapter
```

The following figure shows the list of smali files in the directory:

```
mare@ubuntu:~/Desktop/app/app-debug/smali_classes5/com/example/securestorage/adapter$ ls -l
total 24
-rw-rw-r-- 1 mare mare  2411 Mar 24 00:59 'CardDetailsAdapter$1.smali'
-rw-rw-r-- 1 mare mare  2413 Mar 24 00:59 'CardDetailsAdapter$2.smali'
-rw-rw-r-- 1 mare mare  3632 Mar 24 00:59 'CardDetailsAdapter$CardViewHolder.smali'
-rw-rw-r-- 1 mare mare 10866 Mar 24 00:59  CardDetailsAdapter.smali
```

Figure 3.9 – smali files for classes4.dex

Now, we can read the content of the smali file using the cat utility:

```
# cat CardDetailsAdapter.smali
```

The screenshot for reference is as follows:

```
mare@ubuntu:~/Desktop/app/app-debug/smali_classes5/com/example/securestorage/adapter$ cat CardDetailsAdapter.smali
.class public Lcom/example/securestorage/adapter/CardDetailsAdapter;
.super Landroidx/recyclerview/widget/RecyclerView$Adapter;
.source "CardDetailsAdapter.java"

# annotations
.annotation system Ldalvik/annotation/MemberClasses;
    value = {
        Lcom/example/securestorage/adapter/CardDetailsAdapter$CardViewHolder;
    }
.end annotation

.annotation system Ldalvik/annotation/Signature;
    value = {
        "Landroidx/recyclerview/widget/RecyclerView$Adapter<",
        "Lcom/example/securestorage/adapter/CardDetailsAdapter$CardViewHolder;",
        ">;"
    }
.end annotation

# instance fields
.field private cardInfoList:Ljava/util/List;
    .annotation system Ldalvik/annotation/Signature;
        value = {
            "Ljava/util/List<",
            "Lcom/example/securestorage/model/CardInfo;",
            ">;"
        }
    .end annotation
.end field
```

Figure 3.10 – Reading the smali file for classes4.dex

It would also be a good exercise to compare the code in JADX and `smali` for the same section of the application, `CardDetailsAdapter`.

The following figure shows a comparison between the Java source code obtained using JADX and the respective `smali` code for the same section:

Figure 3.11 – Comparing the code in JADX and smali

Let's summarize what we have done so far:

- Extracted the content of the APK, by using a tool such as unzip. This is not the decompiling of the APK but a simple extraction of contents, following the unarchive process. That's why the compiled resources such as AndroidManifest.xml will not be readable. (Refer to *Figure 3.5.*)

- Used the JADX tool to get the decompiled Java source code from the APK. It is easier to use this decompiled code in JADX to understand the functionalities of the application and different classes, for example. However, if we needed to modify any of the content of this source code, it would be very difficult to recompile it. Also, JADX might not be able to convert all the .dex files properly to Java code (in readable format). (Refer to *Figure 3.6.*)

- Decompiled the APK using apktool, which also resulted in getting the smali version of the code. The smali format is comparatively easy to modify and recompile. (Refer to *Figure 3.7.*)

We can also use the smali/baksmali tools independently to convert the .dex files to smali code, and vice versa. To do so, we can take any of the .dex files from ZIP extracted contents and run baksmali on them.

Let's take the classes4.dex file and copy it to the folder where the smali/baksmali tools are saved:

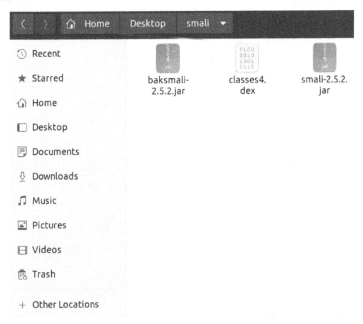

Figure 3.12 – Copying the classes4.dex file to the smali tool folder

Now, open the Terminal window in the same folder by right-clicking and selecting the **Open in Terminal** option. In the Terminal window, run the following command:

```
# java -jar baksmali-2.5.2.jar disassemble classes4.dex -o app
```

You will see a new directory created with the name app and it will have the smali files located at app/com/example/securestorage/adapter:

```
mare@ubuntu:~/Desktop/smali/app/com/example/securestorage/adapter$ ls -l
total 24
-rw-rw-r-- 1 mare mare  2525 Dec  6 02:00 'CardDetailsAdapter$1.smali'
-rw-rw-r-- 1 mare mare  2527 Dec  6 02:00 'CardDetailsAdapter$2.smali'
-rw-rw-r-- 1 mare mare  3635 Dec  6 02:00 'CardDetailsAdapter$CardViewHolder.smali'
-rw-rw-r-- 1 mare mare 10896 Dec  6 02:00  CardDetailsAdapter.smali
```

Figure 3.13 – smali file for classes4.dex

Often, reverse engineering is used to find the solution to a specific question, for example, *How is this application storing the user information?* or *How is the application implementing root detection?* Let's have a look at a similar case and understand when we can use reverse engineering to find security issues during a penetration test.

# Reverse engineering and penetration testing

As we have successfully reverse engineered an APK to the Java source code, it is now important to understand why reverse engineering is very important and might be required during penetration testing. Often in a penetration testing engagement (black box), all that is available to the penetration tester is the name of the application. The penetration tester downloads the application on a device and extracts the APK.

There might be several cases when it is not evident how certain functions of the application are implemented just by using the app. In order to find vulnerabilities in the application, it is required to understand how it works. Reverse engineering helps to answer some of these questions.

Let's take an example. Imagine you are given a banking application to test (penetration testing). While using the application, you notice that the application implements a security control that encrypts all the user-submitted data values before sending them as a part of an HTTP request to the backend server.

For example, a login request (HTTP), when captured, might look as follows:

```
POST api/login HTTP/1.1
HOST: applicationdomainname.com
Content-Type: application/json
```

```
{
        "email":" 41ZEyV2TFKvkjJwulP7I4hY8qEZaYagik2R6BHJFrPg= ",
        "password":"crGTh+mckBpwBxXOKTQpWQ=="
}
```

In this case, it is important to understand how the values of the email and the password are being encrypted to create ciphertext that is being sent as a part of the HTTP request. The answer to this question might be found by reverse engineering the application and exploring the source code to understand the classes where the encryption of input values is being done.

Similarly, there could be other examples in which different functionalities of the app have a complicated implementation. In all such cases, reverse engineering the application generally helps in understanding the logic.

As well as that, there are other cases where the reverse engineering of the application can be very much required during a penetration test. Some of the cases are as follows:

- Finding the API calls, or endpoints that the application is making to the backend.

- Understanding the way some security controls are implemented in order to bypass them. For example, certificate pinning is a security control implemented in a lot of mobile applications to ensure that an application only establishes the TLS connection using the certificate inside the package, and no external user-installed or system-installed TLS certificate is trusted. To implement this, the application code verifies that the TLS certificate presented during the SSL handshake is the same as the one stored inside the application package.

  One of the common tests performed during root detection is to verify whether an application with the name `SuperUser` is installed on the device or not. By reverse engineering the application, a penetration tester can find these types of tests that are being done by the app. Then, they can modify the corresponding `smali` file to return a false result and therefore bypass the root detection.

- Finding hardcoded sensitive information inside the application code, such as backdoor accounts, API keys and secret, unpublished backend endpoints, and hidden menus.

- Finding interesting strings in the code.

- Finding points of encryption and obfuscation, so that they can be decrypted and de-obfuscated.

Reverse engineering also helps in learning more about the important components of an application. A penetration tester would be able to find details of the following components through reverse engineering:

- **Activities**: Components that provide a screen with which users can interact. For example, *Figure 3.2*, *Figure 3.3*, and *Figure 3.4* show the activities of the SecureStorage application.

- **Broadcast receivers**: Components that receive and respond to broadcast messages from other apps or from the operating system.

- **Services**: Components that perform operations in the background.

The majority of these components are listed in the `AndroidMinfest.xml` file, and the same components can be read/explored from the JADX code or `smali` files. Let's look at the `AndroidManifest.xml` file of the SecureStorage app:

```xml
<?xml version="1.0" encoding="utf-8"?>

<manifest xmlns:android="http://schemas.android.com/apk/res/android" android:versionCode="1" android:versionName="1.0" android:compileSdkVersion="30" android:compileSdkVersionCodename="11" package="com.example.securestorage" platformBuildVersionCode="30" platformBuildVersionName="11">

    <uses-sdk android:minSdkVersion="21" android:targetSdkVersion="30"/>

    <application android:theme="@style/Theme.SecureStorage" android:label="@string/app_name" android:icon="@drawable/ic_icon" android:name="com.example.securestorage.activity.MyApplication" android:debuggable="true" android:allowBackup="true" android:supportsRtl="true" android:roundIcon="@drawable/ic_round_icon" android:appComponentFactory="androidx.core.app.CoreComponentFactory">

        <activity android:name="com.example.securestorage.activity.LoginActivity">

            <intent-filter>

                <action android:name="android.intent.action.MAIN"/>

                <category android:name="android.intent.category.LAUNCHER"/>

            </intent-filter>

        </activity>

        <activity android:name="com.example.securestorage.activity.HomeActivity"/>

        <activity android:name="com.example.securestorage.activity.SignupActivity"/>

        <activity android:name="com.example.securestorage.activity.SaveCardInfoActivity"/>

        <activity android:name="com.example.securestorage.activity.CardDetailsActivity"/>

    </application>

</manifest>
```

Figure 3.14 – AndroidManifest.xml file content

The `AndroidManifest.xml` file mentions that there are five activities in this app:

- `com.example.securestorage.activity.LoginActivity`
- `com.example.securestorage.activity.HomeActivity`
- `com.example.securestorage.activity.SignupActivity`
- `com.example.securestorage.activity.SaveCardInfoActivity`
- `com.example.securestorage.activity.CardDetailsActivity`

We can further explore the implementation of any of these activity components as shown in the following figure in the respective section in JADX:

Figure 3.15 – Exploring application activities in JADX

We can select an activity and look at the code. Let's see how `SaveCardInfoActivity` is implemented:

```
package com.example.securestorage.activity;

import android.os.Bundle;
import android.text.Editable;
import android.text.TextUtils;
import android.text.TextWatcher;
import android.view.View;
import android.widget.EditText;
import android.widget.TextView;
import androidx.appcompat.app.AppCompatActivity;
import com.example.securestorage.R;
import com.example.securestorage.model.CardInfo;
import com.example.securestorage.utils.CommonUtils;
import com.example.securestorage.utils.Constants;
import com.example.securestorage.utils.SaveDataUtils;
import java.text.ParseException;
import java.text.SimpleDateFormat;
import java.util.Calendar;
import java.util.Locale;

public class SaveCardInfoActivity extends AppCompatActivity {
    private EditText etCVV;
    private EditText etCardHolderName;
    private EditText etCardNumber;
    private EditText etExpirationDate;
    private boolean isInEditMode;
    private int itemPosition;
    private String mLastInput;

    /* access modifiers changed from: protected */
    @Override // androidx.activity.ComponentActivity, androidx.core.app.ComponentActivity, androidx.fragment.app.FragmentActivity
    public void onCreate(Bundle savedInstanceState) {
        super.onCreate(savedInstanceState);
        setContentView(R.layout.activity_save_card_info);
        this.etCardNumber = (EditText) findViewById(R.id.et_card_number);
        this.etExpirationDate = (EditText) findViewById(R.id.et_expiration_date);
        this.etCVV = (EditText) findViewById(R.id.et_cvv);
        this.etCardHolderName = (EditText) findViewById(R.id.et_card_owner_name);
        Bundle bundle = getIntent().getExtras();
        if (bundle != null) {
            this.isInEditMode = bundle.getBoolean(Constants.IS_EDIT_CARD_DETAILS);
            int i = bundle.getInt(Constants.ITEM_POSITION);
            this.itemPosition = i;
            if (this.isInEditMode) {
                CardInfo cardInfoAtPosition = SaveDataUtils.getCardInfoAtPosition(i);
                this.etCardNumber.setText(SaveDataUtils.getDecryptedData(cardInfoAtPosition.getCardNumber()));
                this.etExpirationDate.setText(SaveDataUtils.getDecryptedData(cardInfoAtPosition.getCardExpiry()));
                this.etCVV.setText(SaveDataUtils.getDecryptedData(cardInfoAtPosition.getCardCvv()));
```

Figure 3.16 – SaveCardInfoActivity

Interestingly, if you look at the following section of code, you can see that the application performs some kind of encryption before saving the card details on the device. It might be interesting to find out how the application encrypts the data submitted and if possible, find a weakness in that section:

```
    private void updateCardInfo() {

        CardInfo cardInfo = new CardInfo();

        cardInfo.setCardNumber(SaveDataUtils.
getEncryptedData(this.etCardNumber.getText().toString()));

        cardInfo.setCardExpiry(SaveDataUtils.
getEncryptedData(this.etExpirationDate.getText().toString()));

        cardInfo.setCardCvv(SaveDataUtils.
getEncryptedData(this.etCVV.getText().toString()));
```

```
        cardInfo.setCardHolderName(SaveDataUtils.
getEncryptedData(this.etCardHolderName.getText().toString()));
        SaveDataUtils.updateCardInfo(cardInfo, this.
itemPosition);
        CommonUtils.showToast(this, "Card Details Edit
Successfully");
    }


    /* access modifiers changed from: private */
    /* access modifiers changed from: public */
    private void saveCardInfo() {
        SaveDataUtils.addCardInfo(this.etCardNumber.getText().
toString(), this.etExpirationDate.getText().toString(), this.
etCVV.getText().toString(), this.etCardHolderName.getText().
toString());
        this.etCardNumber.setText("");
        this.etExpirationDate.setText("");
        this.etCVV.setText("");
        this.etCardHolderName.setText("");
        CommonUtils.showToast(this, "Card Details Saved
Successfully");
    }
```

It should also be noted that there is an EncrytionUtils inside the utils section. EncryptionUtils is called via the getEncryptedData function (inside SaveDataUtils):

```
public class EncryptionUtils {
    public static final String password = "qkjll5@2md3gs5Q@";

    public static SecretKey generateKey() {
        return new SecretKeySpec(password.getBytes(), "AES");
    }

    public static byte[] encryptMsg(String message)
throws NoSuchAlgorithmException, NoSuchPaddingException,
InvalidKeyException, IllegalBlockSizeException,
BadPaddingException, UnsupportedEncodingException {
```

```
        SecretKey secret = generateKey();
        Cipher cipher = Cipher.getInstance("AES/ECB/
PKCS5Padding");
        cipher.init(1, secret);
        return cipher.doFinal(message.getBytes("UTF-8"));
    }


    public static String decryptMsg(byte[] cipherText)
throws NoSuchPaddingException, NoSuchAlgorithmException,
InvalidKeyException, BadPaddingException,
IllegalBlockSizeException, UnsupportedEncodingException {
        SecretKey secret = generateKey();
        Cipher cipher = Cipher.getInstance("AES/ECB/
PKCS5Padding");
        cipher.init(2, secret);
        return new String(cipher.doFinal(cipherText), "UTF-8");
    }
}
```

From the preceding code, we can see that the application seems to be performing **Advanced Encryption Standard** (**AES**) encryption on the data before saving it on the device. The encryption is a symmetric encryption, which uses the same key to encrypt and decrypt the data. The key is also a part of the decompiled source code.

That's a security issue – hardcoding the encryption/decryption key in the application code itself. The key is mentioned in the following line:

```
public static final String password = "qkjll5@2md3gs5Q@";
```

There are several different types of vulnerability that can be discovered by analyzing the reverse engineered code. For example, you can find arbitrary code execution if a section of application code allows code from other apps to run. This type of issue could be discovered by reverse engineering the application and analyzing its code.

So far, we have been working on a debug release of the application. A debug release does not always contain all the security controls that the release build of the application has. One of the most important things missing in the debug release is often code obfuscation.

# Modifying and recompiling the application

Often, it is necessary to not just reverse engineer the application, but also change something, and then repack it. To create a modified APK, you will need to recompile the modified code and then sign the APK again.

Let's say we want to modify the encryption key in the application and then recompile it. To do so, you will need to perform the following steps:

1. Decompile the APK. We have already decompiled the SecureStorage application using `apktool`, with the `#apktool d app-debug.apk` command.

2. The decompilation process will provide us with the required `smali` files. So, let's open the `EncryptionUtils.smali` file.

3. In this `smali` file, we can change the value of the encryption key to something else, such as `abcdef12345`.

4. To recompile the application, we can again use `apktool`. Run the `#apktool b` command. Ensure that this command runs in the same directory where the application was extracted. It will compile the new APK inside the `dist` folder.

> **Important Note**
> To install this modified APK on a device, we will need to sign in with a key. You can generate a key using the `keytool` tool and then use `jarsigner` to sign the application again.

Making smaller changes, such as modifying strings or changing a few static elements of the application, is easier and can be done simply by following the preceding steps. However, to recompile the application after big changes in the `smali` code, you will need to follow some more recommendations.

# Code obfuscation in Android apps

Code obfuscation is a process of modifying the code to protect intellectual property and to make it difficult to reverse engineer. Code obfuscation only modifies the method instructions or metadata; it does not change the logic/flow or the output of the code operation.

Android malware is also known to utilize obfuscation techniques to hide its malicious behavior. However, obfuscation can also be defeated. A skilled reverse engineer would be able to defy the obfuscation techniques implemented and still find the interesting bits in the application code.

Developers may use the default obfuscation tool **ProGuard**, available in Android Studio, or also use a third-party obfuscation tool available in the market. Depending upon the type of obfuscation used, the de-obfuscation technique should be changed. ProGuard is an open source command-line tool that can be used to obfuscate Java code.

One of the ways to de-obfuscate the DEX bytecode is by identifying and using the de-obfuscation methods in the application. This can be done by running the Java code of de-obfuscation methods (the de-obfuscation method code implemented) on the other classes that you want to de-obfuscate.

To understand better, download the release build of the SecureStorage application from the following link:

- SecureStorage (Release Build): `https://github.com/0ctac0der/` `SecureStorage-AndroidApp/releases/download/1.0/` `app-release.apk`. This release build has a basic level of obfuscation implemented on it using ProGuard. Let's decompile it using JADX. Follow the same steps as followed in the *Extracting the Java source code* section.

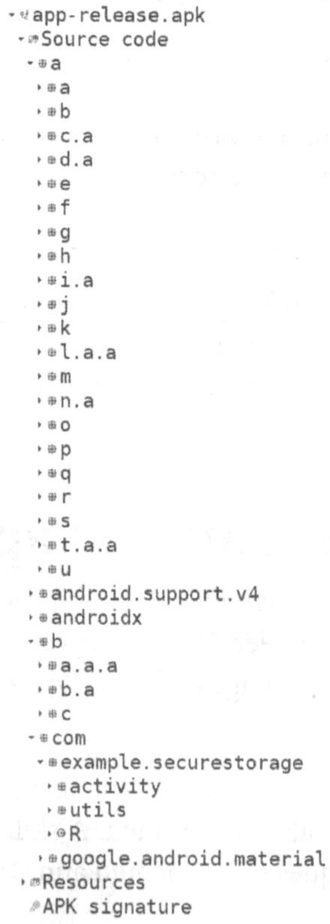

```
- app-release.apk
 - Source code
  - a
   - a
   - b
   - c.a
   - d.a
   - e
   - f
   - g
   - h
   - i.a
   - j
   - k
   - l.a.a
   - m
   - n.a
   - o
   - p
   - q
   - r
   - s
   - t.a.a
   - u
  - android.support.v4
  - androidx
  - b
   - a.a.a
   - b.a
   - c
  - com
   - example.securestorage
    - activity
    - utils
    - R
   - google.android.material
  - Resources
  - APK signature
```

Figure 3.17 – ProGuard obfuscation

You can see that the class names have been modified to random letters. On further analysis, you will notice that the `utils` section no longer seems to have classes other than `CommonUtils`. But, for the app to function, the encryption class and the key have to be there in the code itself.

It is possible to further explore the reverse engineered source code to find the correct place where the `EncryptionUtils` class is:

Figure 3.18 – Obfuscated code of the encryption class

We can note that the encryption key being used in the application is still the same, even after the code has been obfuscated. This is because obfuscation is performed in such a way that the functioning of the application is not changed at all.

The preceding obfuscated code is a result of ProGuard obfuscation, based on the following rule:

```
# class:
#-keepclassmembers class fqcn.of.javascript.interface.for.
webview {
#    public *;
```

```
#}


# Uncomment this to preserve the line number information for
# debugging stack traces.
#-keepattributes SourceFile,LineNumberTable


# If you keep the line number information, uncomment this to
# hide the original source file name.
#-renamesourcefileattribute SourceFile


-keep class com.example.securestorage.utils.CommonUtils { *; }
```

The ProGuard rule states that com.example.securestorage.utils.
CommonUtils should be kept as it is, and the rest of the application code should be obfuscated. This is exactly what you see in the JADX decompiled code for the release build of the application.

While performing penetration testing, it is not always necessary to de-obfuscate the whole code. Often, you will only need to understand the code logic, or just de-obfuscate some part of the code. There are also de-obfuscation tools available, which sometimes can be useful if you are really stuck with some part of the application code, although I would like to recommend a manual analysis of the code to understand it better; only use a de-obfuscation tool as the last resort.

## Summary

This chapter explained how Android applications are developed, compiled, and packed. We learned how to perform reverse engineering on Android applications to create the original Java source code. Once the Java source code is decompiled from the APK, we learned what to look for in the app and how to find security issues. Obfuscation and de-obfuscation are also important parts of reverse engineering, and we learned how a developer may implement some basic ProGuard obfuscation on the application code before creating the release build. However, it is not always required that the whole decompiled application code is de-obfuscated as well.

In the next chapter, we are going to have a closer look at reverse engineering an iOS application. We will explore the tools used for that and will also learn how to enumerate interesting bits in the decompiled application binary of an iOS application.

# 4
# Reverse Engineering an iOS Application

In comparison to Android apps, reverse engineering an iOS application is a bit more complicated. This is mainly due to the security controls that are implemented by iOS and the way Apple manages application installation and verification across all iOS devices. For example, to get the **iOS App Store Package (IPA)** from a device running on the application, you can't simply extract the IPA and install it on another iOS device. This is because all the applications that are installed from the Apple App Store are encrypted on the device. Here, you would be required to extract a decrypted IPA and then sign it again to be able to run it on another iOS device.

In this chapter, we will cover the following topics:

- Learning more about how iOS apps are developed
- Understanding the iOS executable format
- Exploring more about iOS app reverse engineering tools and their usage
- Reverse engineering the SecureStorage iOS application

# Technical requirements

We will be using the same virtual machine setup for Ubuntu that we used in the previous chapter. However, it might be useful if you have a Mac laptop or computer and an iOS device.

For this chapter, the following tools need to be installed:

- Hopper Disassembler (`https://www.hopperapp.com/` – the free version should be good to get started with)

- Ghidra (`https://ghidra-sre.org/`)

# iOS app development

iOS apps are commonly developed using the Swift or Objective-C languages. Objective-C is a general-purpose programming language with object-oriented capabilities and a dynamic runtime. Until 2014, Objective-C was the official language for iOS app development.

Apple launched Swift in 2014, a general-purpose, high-level programming language designed to develop apps for Apple's operating systems. Initially, it was a proprietary language, but version 2.2 was made open source under Apache License 2.0.

For iOS application development, Xcode is the official **integrated development environment** (**IDE**). Developers also have the option of choosing other IDEs, such as AppCode or Visual Studio Code from Microsoft, but these IDEs also need Xcode underneath to work properly. Xcode includes the required **software development kits** (**SDKs**), tools, compilers APIs, and so on. Xcode uses the `swiftc` compiler for Swift and the `clang` compiler for Objective-C code.

The following diagram illustrates how the source code (in Swift) is processed and compiled using Xcode's build system:

Figure 4.1 – Xcode's build system

Once the compilation process has been completed, the IDE creates the final package – that is, the IPA. The IPA contains all the necessary resources, certificates, assets, and binary files. Before we start reversing, it is important to learn about the structure of the executable binary.

# Understanding the binary format

For systems that are based on the Mach Kernel, such as macOS and iOS, Mach-O is the format that's used for the executable files and shared libraries. **Mach-O** stands for **Mach object file format**. The applications are expected to run on different processor types, for the most part. Owing to this, the executable code should be native to different instruction sets.

Depending on the instruction sets it contains, a binary file is called a thin binary if it contains a single executable file for one architecture; it's called a fat binary if it contains code for different CPU instruction sets in a single file – that is, it has been fattened (or expanded).

Each binary file begins with a header (called a mach header) that contains a magic number that can be used to identify it. For a thin binary file, the header contains one magic number; however, for a fat binary, the header is a fat header. The fat header contains the locations of the mach headers of the other executables in it.

| Thin Binary | Fat Binary |
| --- | --- |
| Contains machine code for one architecture type | Contains machine code for multiple architecture types |
| Smaller in size than fat binaries | Bigger in size than thin binaries |

As we dive deeper into the binary structure and reverse it, we'll use a sample application and perform the same steps. Here, we will be using the iOS version of the **SecureStorage** application. The iOS application has two variants – one that's developed using Objective-C and another that's been developed using Swift. In this chapter, we will be working with the Objective-C version of the app.

You can download the Objective-C version of the application from the following GitHub link:

- The SecureStorage (Objective-C version) application package (`https://github.com/0ctac0der/SecureStorage-ObjC/releases/download/0.1/SecureStorageObjC.ipa`)

IDEs for iOS apps in Xcode can only run on Mac-based systems. However, since we are working on reverse engineering the application, we do not need the Xcode IDE. The only thing we need to start is the application package – that is, the IPA – and this can be downloaded from the link mentioned previously. The iOS version of the application also looks the same as the Android version. The following are some of the screenshots of the SecureStorage application running on iOS:

Figure 4.2 – The SecureStorage application (iOS version)

With that, we have discussed the basics of binary formats and have an understanding of the SecureStorage app as well. Now, let's learn about the process of reverse engineering these apps and the binaries inside them.

# Reverse engineering an iOS app

In a "black box" penetration test, it is the job of the penetration tester to somehow extract the application package (IPA) from a device. When an application is installed from the Apple App Store, it is protected by **Digital Rights Management** (**DRM**). The application binary file is encrypted when it is stored on the iOS device. That's why simply extracting the binary from the device and reverse engineering it is not going to work.

Extracting an unencrypted application from an iOS device can be done using tools such as frida-ios-dump (https://github.com/AloneMonkey/frida-ios-dump) and frida-ipa-dump (https://github.com/integrity-sa/frida-ipa-dump). The steps to extract the unencrypted IPA won't be covered in this book since that is part of the penetration testing process. However, to learn more about how to extract the unencrypted IPA, it is recommended to follow the steps given on the GitHub page for frida-ios-dump (or frida-ipa-dump).

> **Important Note**
>
> To extract the unencrypted IPA, we need a jailbroken iOS device running the target application. When the app runs in the device memory (process memory), it is in an unencrypted state. Utilities such as `frida-ios-dump` and `frida-ipa-dump` extract all the unencrypted application files from memory and combine them in the IPA. Reverse engineering can then be performed on the extracted IPA, but to reinstall that IPA on another device, it would need to be signed by an Apple developer certificate. Another utility (`https://www.iosappsigner.com/`) can be used to sign the IPA.

For now, let's download the package from the GitHub link mentioned previously (`https://github.com/0ctac0der/SecureStorage-ObjC/releases/download/0.1/SecureStorageObjC.ipa`) and save it in a directory called iOS on our virtual machine environment's desktop. As we already know, the IPA is also a compressed (ZIP) file containing different assets and files. So, let's unzip the file directly using the `unzip` utility in Linux.

```
mare@ubuntu:~/Desktop/ios$ unzip SecureStorageObjC.ipa
Archive:  SecureStorageObjC.ipa
   creating: Payload/
   creating: Payload/SecureStorageObjC.app/
   creating: Payload/SecureStorageObjC.app/_CodeSignature/
  inflating: Payload/SecureStorageObjC.app/_CodeSignature/CodeResources
   creating: Payload/SecureStorageObjC.app/Base.lproj/
   creating: Payload/SecureStorageObjC.app/Base.lproj/Main.storyboardc/
  inflating: Payload/SecureStorageObjC.app/Base.lproj/Main.storyboardc/UINavigationController-aPx-F6-XA7.nib
  inflating: Payload/SecureStorageObjC.app/Base.lproj/Main.storyboardc/CardsViewController.nib
  inflating: Payload/SecureStorageObjC.app/Base.lproj/Main.storyboardc/6aw-hB-w9F-view-ndu-Je-xTM.nib
  inflating: Payload/SecureStorageObjC.app/Base.lproj/Main.storyboardc/LoginViewController.nib
  inflating: Payload/SecureStorageObjC.app/Base.lproj/Main.storyboardc/ProUserPaymentViewController.nib
  inflating: Payload/SecureStorageObjC.app/Base.lproj/Main.storyboardc/AfterLoginViewController.nib
  inflating: Payload/SecureStorageObjC.app/Base.lproj/Main.storyboardc/lXa-7A-rEr-view-ljb-0a-MCS.nib
  inflating: Payload/SecureStorageObjC.app/Base.lproj/Main.storyboardc/MmE-Yw-i35-view-vak-YX-vKG.nib
  inflating: Payload/SecureStorageObjC.app/Base.lproj/Main.storyboardc/Info.plist
  inflating: Payload/SecureStorageObjC.app/Base.lproj/Main.storyboardc/Gbl-Pj-ADR-view-vgt-DE-73h.nib
  inflating: Payload/SecureStorageObjC.app/Base.lproj/Main.storyboardc/EFY-PK-qel-view-T1t-V3-Iqe.nib
  inflating: Payload/SecureStorageObjC.app/Base.lproj/Main.storyboardc/SignUpViewController.nib
   creating: Payload/SecureStorageObjC.app/Base.lproj/LaunchScreen.storyboardc/
  inflating: Payload/SecureStorageObjC.app/Base.lproj/LaunchScreen.storyboardc/01J-lp-oVM-view-Ze5-6b-2t3.nib
  inflating: Payload/SecureStorageObjC.app/Base.lproj/LaunchScreen.storyboardc/UIViewController-01J-lp-oVM.nib
  inflating: Payload/SecureStorageObjC.app/Base.lproj/LaunchScreen.storyboardc/Info.plist
  inflating: Payload/SecureStorageObjC.app/Assets.car
  inflating: Payload/SecureStorageObjC.app/SecureStorageObjC
  inflating: Payload/SecureStorageObjC.app/embedded.mobileprovision
  inflating: Payload/SecureStorageObjC.app/Info.plist
  inflating: Payload/SecureStorageObjC.app/PkgInfo
```

Figure 4.3 – Unzipping the IPA

As shown in the preceding screenshot, once the file has been unzipped, a new folder will be created called `Payload`. All the iOS packages contain a `Payload` folder. This `Payload` folder contains a `.app` file, which often has the same name as the binary (application). In this case, if you navigate inside the `Payload` folder, you will find the `.app` file – that is, `SecureStorageObjC.app`:

```
mare@ubuntu:~/Desktop/ios$ cd Payload/
mare@ubuntu:~/Desktop/ios/Payload$ ls
SecureStorageObjC.app
mare@ubuntu:~/Desktop/ios/Payload$ 
```

Figure 4.4 – Locating the .app file

The `.app` file is shown as a package file on the Mac operating system, and right-clicking it will show the **Show Package Contents** option.

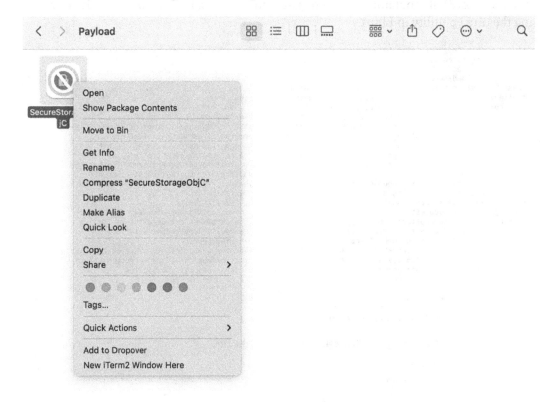

Figure 4.5 – Navigating inside the Payload folder (on Mac)

On our Ubuntu virtual machine, we can simply run the following command to navigate inside the `.app` file.

```
mare@ubuntu:~/Desktop/ios/Payload$ ls
SecureStorageObjC.app
mare@ubuntu:~/Desktop/ios/Payload$ cd SecureStorageObjC.app/
mare@ubuntu:~/Desktop/ios/Payload/SecureStorageObjC.app$ ls -l
total 352
-rw-r--r-- 1 mare mare 160744 Jan  1 21:54 Assets.car
drwxr-xr-x 4 mare mare   4096 Jan  1 21:54 Base.lproj
drwxr-xr-x 2 mare mare   4096 Jan  1 21:54 _CodeSignature
-rw-r--r-- 1 mare mare  17970 Jan  1 21:54 embedded.mobileprovision
-rw-r--r-- 1 mare mare   3734 Jan  1 21:54 Info.plist
-rw-r--r-- 1 mare mare      8 Jan  1 21:54 PkgInfo
-rwxr-xr-x 1 mare mare 157376 Jan  1 21:54 SecureStorageObjC
mare@ubuntu:~/Desktop/ios/Payload/SecureStorageObjC.app$ █
```

Figure 4.6 – Navigating inside the Payload folder (on Linux)

This .app file contains all the assets of the application, such as the images, certificates, and supportive components that are required for the app to work. One of the important files here is the Info.plist file. This is similar to the AndroidManifest.xml file in Android applications. The Info.plist file contains information related to the configuration and components of the application, such as the bundle's name, transport layer security configuration, URL schemes, and MinimumOSVersion.

To understand the application in more depth, we should start by analyzing all these resource files. Some of the interesting files to look into are the .plist files, configuration files, and so on. All these files may contain interesting information.

Along with the supportive files, there is also an executable file with the same name as that of the app. This is the app's binary file.

This executable file is what we will be reverse engineering to extract useful information. We will utilize different tools to reverse engineer the binary.

## Extracting strings from the binary

The first step is to gather information about the different app components. For this purpose, we can use the strings tool to dump all the strings that are hardcoded in the binary.

To run the strings utility on the binary and save all the extracted strings in a separate file (strings.txt), use the following command after navigating inside the app directory:

```
# ls -l (To list all the contents of the directory)
# strings SecureStorageObjC >> strings.txt
```

The following screenshot shows the output of these commands:

```
mare@ubuntu:~/Desktop/ios/Payload/SecureStorageObjC.app$ ls -l
total 352
-rw-r--r-- 1 mare mare 160744 Jan  1 21:54 Assets.car
drwxr-xr-x 4 mare mare   4096 Jan  1 21:54 Base.lproj
drwxr-xr-x 2 mare mare   4096 Jan  1 21:54 _CodeSignature
-rw-r--r-- 1 mare mare  17970 Jan  1 21:54 embedded.mobileprovision
-rw-r--r-- 1 mare mare   3734 Jan  1 21:54 Info.plist
-rw-r--r-- 1 mare mare      8 Jan  1 21:54 PkgInfo
-rwxr-xr-x 1 mare mare 157376 Jan  1 21:54 SecureStorageObjC
mare@ubuntu:~/Desktop/ios/Payload/SecureStorageObjC.app$ strings SecureStorageObjC >>strings.txt
```

Figure 4.7 – Running strings on the binary

Once all the strings have been stored in the `strings.txt` file, we can open the text file to read its contents.

```
mare@ubuntu:~/Desktop/ios/Payload/SecureStorageObjC.app$ strings SecureStorageObjC >>strings.txt
mare@ubuntu:~/Desktop/ios/Payload/SecureStorageObjC.app$ cat strings.txt
__PAGEZERO
__TEXT
__text
__TEXT
__stubs
__TEXT
__objc_methlist
__TEXT
__const
__TEXT
__cstring
__TEXT
__objc_methname
__TEXT
__objc_classname__TEXT
__objc_methtype
__TEXT
__unwind_info
__TEXT
__DATA_CONST
__got
__DATA_CONST
__const
__DATA_CONST
__cfstring
__DATA_CONST
__objc_classlist__DATA_CONST
__objc_catlist
__DATA_CONST
__objc_protolist__DATA_CONST
__objc_imageinfo__DATA_CONST
__DATA
__objc_const
__DATA
__objc_selrefs
__DATA
__objc_classrefs__DATA
__objc_superrefs__DATA
__objc_ivar
__DATA
__objc_data
__DATA
__data
__DATA
__LINKEDIT
/usr/lib/dyld
/System/Library/Frameworks/Security.framework/Security
```

Figure 4.8 – Extracted strings

Analyzing all the strings in the application binary is very important while testing the application for any sensitive hardcoded information inside the application code. Several interesting things might be hardcoded in the application, such as an internal IP address, internal application URL, hardcoded key and secret, encryption keys, and more.

We now have all the strings of the binary. At this point, we can analyze them to find interesting hardcoded details.

While developing the iOS applications, all the application code is compiled into machine code. That is why, to analyze the iOS application's binary, we would need a disassembler. So, let's move on and disassemble the binary using a more advanced reverse engineering tool, Ghidra.

# Disassembling the application binary

In this section, we'll learn how to disassemble the application binary using tools such as Ghidra and Hopper.

## Using Ghidra

In *Chapter 1*, *Basics of Reverse Engineering – Understanding the Structure of Mobile Apps*, you learned how to install Ghidra on our virtual machine environment. Now, let's use the same installation steps and disassemble the binary:

1.  To start Ghidra, open a "terminal" and navigate to the directory where Ghidra was downloaded. This directory should contain a file named ghidraRun.

```
mare@ubuntu:~/Desktop/ghidra_10.0.3_PUBLIC$ ls -l
total 48
drwxr-xr-x 5 mare mare  4096 Sep  8 13:19 docs
drwxr-xr-x 5 mare mare  4096 Sep  8 13:19 Extensions
drwxr-xr-x 9 mare mare  4096 Sep  8 13:19 Ghidra
-rwxr-xr-x 1 mare mare   883 Sep  8 13:19 ghidraRun
-rw-r--r-- 1 mare mare   384 Sep  8 13:19 ghidraRun.bat
drwxr-xr-x 6 mare mare  4096 Sep  8 13:19 GPL
-rw-r--r-- 1 mare mare 11357 Sep  8 13:19 LICENSE
drwxr-xr-x 2 mare mare  4096 Sep  8 13:19 licenses
drwxr-xr-x 2 mare mare  4096 Sep  8 13:19 server
drwxr-xr-x 2 mare mare  4096 Sep  8 13:19 support
```

Figure 4.9 – Ghidra directory files

2. To run it, simply run the following command in the Terminal window:

```
# ./ghidraRun
```

3. This will open the Ghidra project window. Choose the **Tools | CodeBrowser** option to open the code browser utility. (Refer to *Chapter 2, Figure 2.5.1 – Starting Ghidra CodeBrowser*.)

4. In the CodeBrowser window, you can drag and drop the binary file to start disassembling it. When the binary file has been dropped in the CodeBrowser window, it will automatically find the format and architecture of the binary.

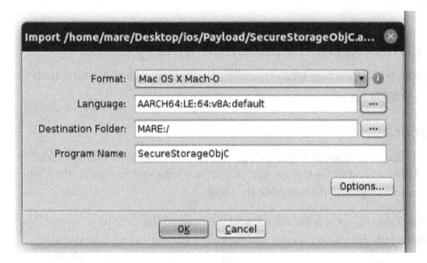

Figure 4.10 – Disassembling a binary using Ghidra CodeBrowser

5. Once you click **OK** in the CodeBrowser window, the binary will be imported and disassembled. While this is being done, the Ghidra tool may also prompt you with an automated analysis option, where it will analyze the file for opcodes and other values.

> **Important Note**
> **Operation code (opcode)** is also known as instruction code/syllable/parcel or opstring code. It is the section of a machine language instruction that explains what operation is to be performed.

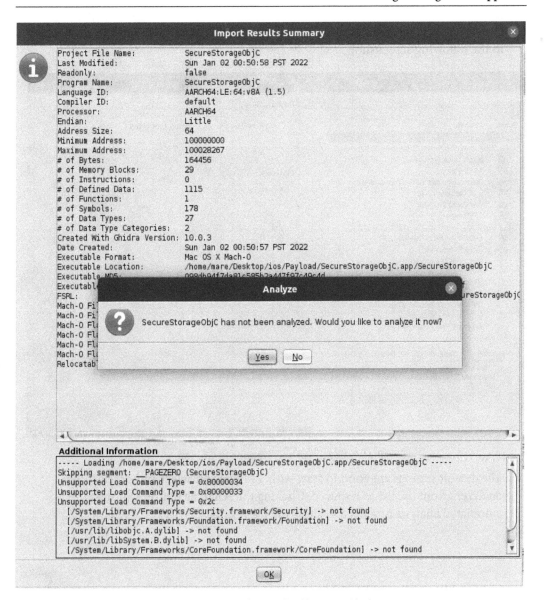

Figure 4.11 – Analyzing the file using Ghidra

6.    Once you click **Yes**, Ghidra will show all the available **Analysis Options**, as shown in the following screenshot:

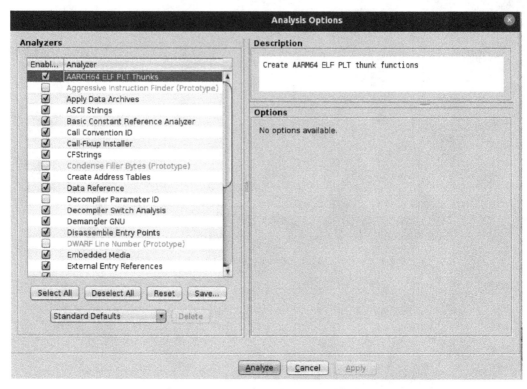

Figure 4.12 – Analysis Options in Ghidra

7.    The default settings are good to start with, but you can select and deselect the analyzers from the list as required. Clicking the **Analyze** button will start the automated analysis process and show the file results.

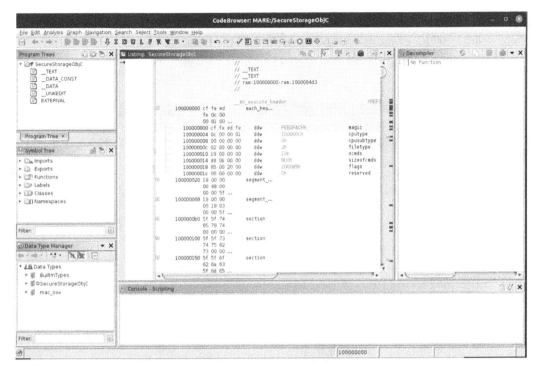

Figure 4.13 – Disassembled binary

**Important Note**

The auto analysis process may show some warnings at the end. Generally, these warnings can be ignored as they are not interesting from a reverse engineering perspective.

Previously in this chapter, we discussed that *each binary file begins with a header (called a mach header), which contains a magic number that can be used to identify it*. The mach header for this binary is visible on the pane in the center, as shown in the following screenshot:

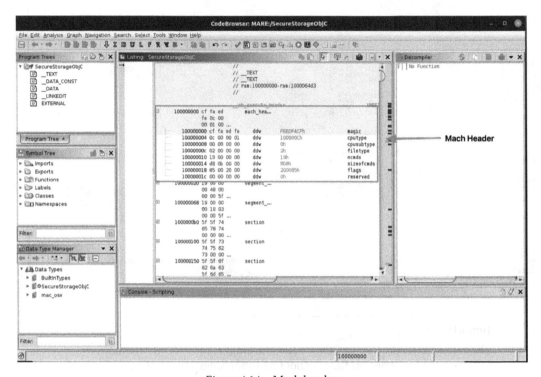

Figure 4.14 – Mach header

The mach header contains load commands that have an 8-byte structure and specifies the logical structure of the file, paths to linked libraries, the code's structure, and the layout of the file in memory.

In the disassembled binary, we can now navigate to various points and sections. For example, let's navigate to **Functions | entry**. This is the first function, which is called at runtime.

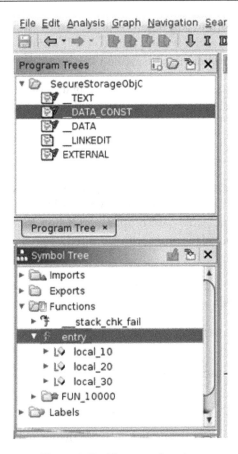

Figure 4.15 – The entry function

When the entry function is selected, Ghidra automatically shows us some approximate code (pseudocode). It is easier to read this code and understand the functionality of the application.

During reverse engineering, you are either searching for an answer to a specific question, such as what type of encryption is being used, or you are interested in understanding how the application is working in the background. Once you have successfully extracted the application package and strings, you will be able to get a lot of information about the application. With the disassembled binary in a tool such as Ghidra, you can navigate across different functions, labels, and so on, and then read the pseudocode to figure out how the app is working. Ghidra is also useful in obtaining cross-references; simply right-click the desired function and select **Show References**.

For the SecureStorage application, you might be interested in finding whether the application encrypts the data, and if it does, then you will want to find out what type of encryption is being used and where the encryption key is.

So far, we have reverse engineered the application using different tools and analyzed it. It is also important to understand how to manually review the disassembled application binaries to find what you are looking for.

## Using Hopper Disassembler

Ghidra is a great tool for reverse engineering iOS applications (and a lot of other binaries too). Hopper Disassembler is another tool that can be used for disassembling iOS app binaries. It can be used to disassemble, decompile, and debug applications. Hopper Disassembler is a paid tool but also comes with a limited demo license with some limitations.

Using Hopper over Ghidra is a personal choice. For example, I like the interface of Hopper a lot more than Ghidra. Installing Hopper Disassembler is very straightforward:

1.  Go to the Hopper Disassembler website at `https://www.hopperapp.com/index.html`.

2.  Click the **Try It Free** button to go to `https://www.hopperapp.com/download.html`.

3.  Download the `.deb` file in the **Linux** section.

4.  Once the `.deb` file has been downloaded inside the virtual machine, run the following command to install it:

    ```
    # sudo dpkg -i Hopper-v4-5.3.0-Linux-demo.deb
    (Replace the deb file name with the correct one you have)
    ```

5.  Once installed, run the application from the `application` directory.

When Hopper Disassembler is run for the first time, it will ask for a **License File**, but as we are only going to use the demo, click the **Try the Demo** button.

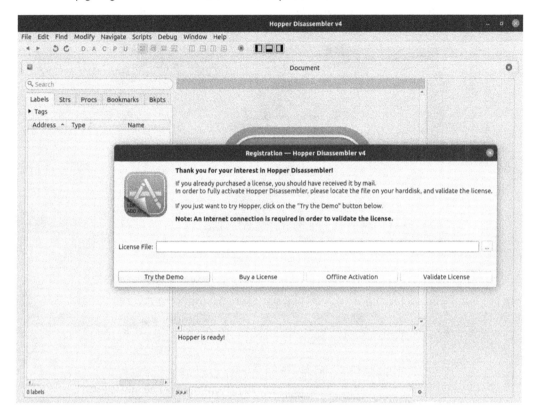

Figure 4.16 – Running Hopper Disassembler

After that, simply drag and drop the application binary into the Hopper Disassembler's middle pane. Similar to Ghidra, Hopper will also provide the details of the binary it is importing. Click on **OK**, without changing any of the other options for checks (unless you are sure about what you are doing).

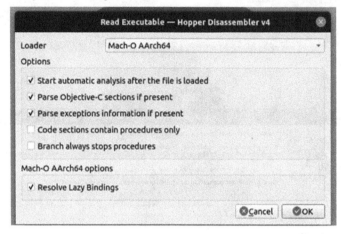

Figure 4.17 – Importing the binary in Hopper Disassembler

Once the binary has been imported, you will see a screen similar to the following:

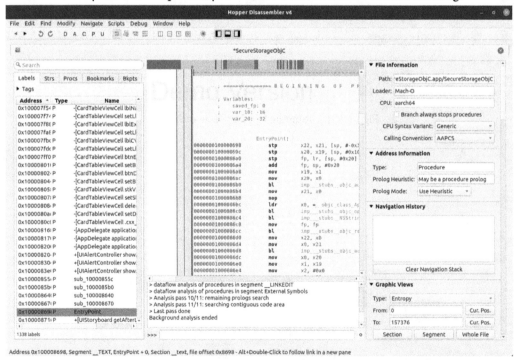

Figure 4.18 – Disassembled binary in Hopper

The tabs on the left pane of Hopper are interesting to start with. The most important tabs are **Labels**, **Strs**, and **Procs**. The **Labels** tab contains different memory addresses, along with their associated names and instruction sets. The **Strs** tab shows all the strings in the binary (similar to what we extracted using the strings utility). All the methods that were used in the application can be found in the **Procs** tab. Hopper also shows the pseudocode for a specific part/function when it's selected. To view the pseudocode for any section, select the section and then click on the **Show Pseudo Code of Procedure** button on the toolbar. The following screenshot shows the pseudocode for the +[**GeneralUtilssave:**] label:

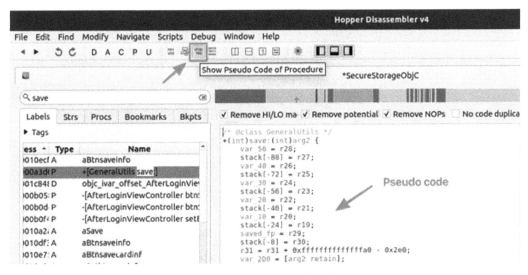

Figure 4.19 – Pseudocode for a label

So far, we have reverse engineered the application using different tools and analyzed it. It is also important to understand how to manually review the disassembled application binaries to find what you are looking for.

# Manually reviewing the disassembled binary for security issues

So far, we have disassembled the binary using Gidra as well as Hopper, and we've also extracted a lot of interesting information from the application package. During a penetration test, the reverse engineered code is manually reviewed to find any security issues. One of the interesting things with regard to the SecureStorage app is finding the type of encryption that's being done and, if possible, the encryption key.

Let's search the **Strs** tab in the disassembled app in Hopper for anything related to the `encrypt` keyword:

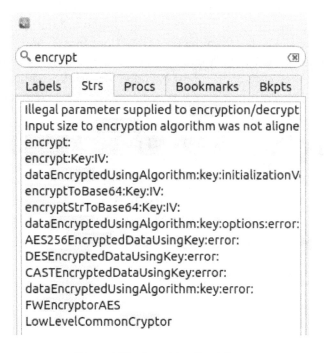

Figure 4.20 – Searching the strings

Out of all the results, the `encrypt:Key:IV:` string looks interesting. We can also inspect all the references to and from this string that are made by the application. To do so, select the string and, on the middle pane, right-click the string and chose the **References to "aEncryptKeyiv"** option:

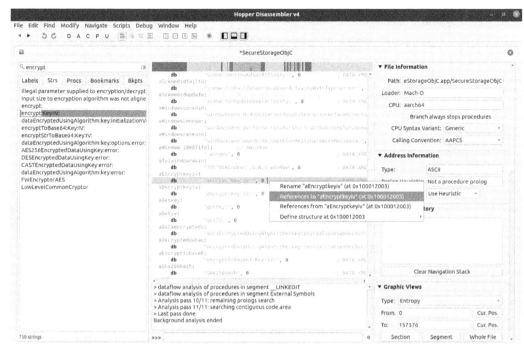

Figure 4.21 – Inspecting all the references

This will result in Hopper showing us all the references to the aEncryptKeyiv keyword. Select the first reference from the list.

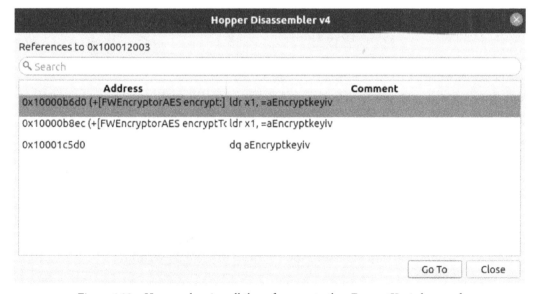

Figure 4.22 – Hopper showing all the references to the aEncryptKeyiv keyword

Looking closely at the selected reference, you will notice a long integer right above it, as shown in the following screenshot. This is the encryption key that's being used.

```
INNING   OF   PROCEDURE ================

rAES encrypt:]:
                                    ; Objective C Implementation defined at 0x10000eaa8 (class method)
   x0, =__objc_class_FWEncryptorAES_class  ; argument "instance" for method imp___stubs__objc_msgSend, __objc_class

   x3, #0x100014510                 ;@"2D4A614E635266556A586E3272357538782F413F4428472B4B6250655367566B"

   x1, =aEncryptkeyiv               ; argument "selector" for method imp___stubs__objc_msgSend, "encryptKe
   x4, x3
   imp___stubs__objc_msgSend        ; objc_msgSend
                                                                                Encryption key
```

Figure 4.23 – The encryption key in the disassembled code

As this is a static hardcoded key inside the application code, it should also be part of the extracted strings. Let's verify the same by using the following command on the secrets.txt file (we are reading through all the secrets and grepping for a part of the encryption key).

```
#cat secrets.txt | grep 2D4A614E6352665
```

You will find that the encryption key was part of the originally extracted secrets. However, it might be difficult to figure out what this key is being used for if we did not have the disassembled binary to analyze.

```
mare@ubuntu:~/Desktop/ios/Payload/SecureStorageObjC.app$ cat strings.txt | grep 2D4A614E6352665
2D4A614E635266556A586E3272357538782F413F4428472B4B6250655367566B
mare@ubuntu:~/Desktop/ios/Payload/SecureStorageObjC.app$ █
```

Figure 4.24 – The encryption key in the string

This was just one example of how reverse engineering the iOS application binary can help you find security issues, as well as understand the functionalities of the application.

To further analyze the application's disassembled code, we must understand the Objective-C runtime. It is also helpful to analyze the disassembled native code, but that would also require having a good understanding of the calling conventions and instructions that are used by the underlying platform. To learn that, you must understand the ARM architecture and ARM assembly basics.

Some Mac-specific tools can be used during the reverse engineering process to provide you with useful information. Let's have a look at these in the next section.

# Using Mac-only tools for iOS app reverse engineering

Along with the tools we have used to reverse engineer the application binary, there are some other useful tools available that primarily work on Mac systems only. Let's have a look at those tools and how they can help in the reverse engineering process:

- **otool**: This is used to dump different libraries and find any deprecated libraries that are being used. The following command can be used to list all the libraries being used in the SecureStorage application:

```
#otool -L [path to binary file]
```

The following screenshot shows the output of otool for the SecureStorage app:

```
0ctac0der@Mac    ~/Desktop/iOS       otool -L SecureStorageObjC
SecureStorageObjC:
        /System/Library/Frameworks/Security.framework/Security (compatibility version 1.0.0, current version 60157.60.19)
        /System/Library/Frameworks/Foundation.framework/Foundation (compatibility version 300.0.0, current version 1856.105.0)
        /usr/lib/libobjc.A.dylib (compatibility version 1.0.0, current version 228.0.0)
        /usr/lib/libSystem.B.dylib (compatibility version 1.0.0, current version 1311.0.0)
        /System/Library/Frameworks/CoreFoundation.framework/CoreFoundation (compatibility version 150.0.0, current version 1856.105.0)
        /System/Library/Frameworks/UIKit.framework/UIKit (compatibility version 1.0.0, current version 5205.0.101)
```

Figure 4.25 – Finding libraries with otool

- **codesign**: This tool is used to analyze signatures, verify entitlements, and so on. One of the common uses of codesign in reverse engineering is to find app entitlements and configurations. To dump the entitlements of the SecureStorage app, run the following command:

```
#codesign -d --entitlements - [path to .app file]
```

The following screenshot shows the dumped entitlements for the SecureStorage app when using codesign:

```
0ctac0der@Mac    ~/Desktop/iOS/Payload      codesign -d --entitlements - SecureStorageObjC.app
Executable=/Users/0ctac0der/Desktop/iOS/Payload/SecureStorageObjC.app/SecureStorageObjC
[Dict]
        [Key] application-identifier
        [Value]
                [String] K824P73V39.com.SecureStorageObjC
        [Key] com.apple.developer.team-identifier
        [Value]
                [String] K824P73V39
        [Key] get-task-allow
        [Value]
                [Bool] true
        [Key] keychain-access-groups
        [Value]
                [Array]
                        [String] K824P73V39.com.SecureStorageObjC
```

Figure 4.26 – Dumping entitlements

Having access to a Mac and being able to run Mac utilities while you're reverse engineering iOS apps can be helpful. However, tools such as strings, Hopper, and Ghidra allow Linux users to perform almost all the necessary tasks.

## Summary

In this chapter, you learned how iOS applications are developed, compiled, and packed. Then, you learned how to reverse engineer iOS apps and disassemble the binary. The reverse engineering process can go even further and help you analyze the disassembled native code and libraries. Reviewing the disassembled or reverse engineered code requires knowing the relevant architecture and instructions. During a penetration test, reverse engineering can help you find several security issues.

In the next chapter, we are going to look at reverse engineering an iOS application that's been developed in Swift. We will look at the differences between reverse engineering a Swift-based application and an Objective-C application.

# 5

# Reverse Engineering an iOS Application (Developed Using Swift)

In the last chapter, we discussed how iOS applications, developed using the Objective C programming language, can be reverse engineered and disassembled to extract useful information from the binary.

Swift is now (at the time of writing – December 2021) the official language for iOS application development. The majority of the modern iOS applications are developed using Swift, as it provides some great benefits, such as better code readability. Swift generates a compiler error as you write code so that issues can be fixed right away, and it has better memory management, less code (in comparison with Objective C), support for dynamic libraries, and so on.

In this chapter, we will be covering the following topics:

- Understanding the difference between Objective C and Swift applications
- Reverse engineering a Swift application

Let's dive in!

# Technical requirements

We will be using the Ubuntu virtual machine setup, which we used in the previous chapter. This chapter will need the following tools to be installed:

- Radare2 (`https://github.com/radareorg/radare2`)
- Ghidra (`https://ghidra-sre.org/`)

# Understanding the difference between Objective C and Swift applications

Swift is a static, strongly typed, high-level programming language that was introduced in 2014. Before Swift, Objective C was primarily the programming language for iOS application development.

For now, you can download the full source code of our Swift application (SecureStorage) or the application package from the following links:

- The GitHub repository (for full source code) of SecureStorage application (the Swift version): `https://github.com/0ctac0der/SecureStorage-iOSApp`.
- The link to download the application package (the SecureStorage Swift version): `https://github.com/0ctac0der/SecureStorage-iOSApp/releases/download/0.1/SecureStorage.ipa`.

Let's have a look at how the code looks for the same application (SecureStorage) in Swift and Objective C, respectively:

```swift
import UIKit

class LoginViewController: UIViewController {

    @IBOutlet weak var txtFEmail: UITextField!{
        didSet{
            self.txtFEmail.roundCorner4Px()
        }
    }
    @IBOutlet weak var txtFPassword: UITextField!{
        didSet{
            self.txtFPassword.roundCorner4Px()
        }|
    }
    @IBOutlet weak var btnLogin: UIButton!{
        didSet{
            self.btnLogin.roundCorner4Px()
        }
    }

    @IBOutlet weak var btnSignUpNow: UIButton!

    var tapCount : Int = 0


    @IBAction func btnLoginClicked(_ sender: Any?) {

        if self.txtFEmail.text?.trimmingCharacters(in: .whitespacesAndNewlines).count ?? 0 == 0 ||
            self.txtFPassword.text?.trimmingCharacters(in: .whitespacesAndNewlines).count ?? 0 == 0{
            self.presentAlert(withTitle: "Error", message: "Please enter user name and password!")
        } else {
            let password = self.txtFPassword.text ?? "XX"
            let userName = self.txtFEmail.text ?? "X"
            if password == Utility.getPassword(forUserName: userName){
                Utility.setPassword(password: password, forUserName: userName)
                Utility.setUserName(userName)
```

Figure 5.1 – The Swift code of the SecureStorage app (LoginViewController)

This is the Objective C code of the SecureStorage app:

```objc
- (void)viewWillAppear:(BOOL)animated {
    [super viewWillAppear:animated];
    [self registerForKeyboardNotifications];
}

- (void)viewWillDisappear:(BOOL)animated {
    [super viewWillDisappear:animated];
    [self deregisterFromKeyboardNotifications];
}


- (IBAction)btnLoginClicked: (id)sender {

    NSString *email = self.txtFEmail.text;
    NSString *password = self.txtFPassword.text;
    if (email.length == 0) {
        [UIAlertController showAlertWithMessage:@"Please enter username" onViewController: self];
        return;
    }
    if (password.length == 0) {
        [UIAlertController showAlertWithMessage:@"Please enter password" onViewController: self];
        return;
    }
    [self loginWithemail:email andPassword:password];
}

- (IBAction)btnSignUpNowClicked: (id)sender {
    SignUpViewController *vc = [UIStoryboard getSignUpViewController];
    [self.navigationController pushViewController:vc animated:YES];
}

- (void)loginWithemail: (NSString *)email andPassword : (NSString *)password {
    NSString* userName = self.txtFEmail.text;
    NSString* key = [userName stringByAppendingString:@"Password"];
    NSString* oldPassword = [[NSUserDefaults standardUserDefaults] objectForKey:key];
    if([oldPassword isEqualToString:self.txtFPassword.text]){
```

Figure 5.2 – The Objective C code of the SecureStorage app (LoginViewController)

In both screenshots (*Figure 5.1* and *Figure 5.2*), some of the visible differences are as follows:

- For each class, the Objective C project (*Figure 5.2*) has a header (a .h file) and an implementation file (a .m file), also known as a compilation unit. However, in Swift, the header is not mentioned separately.

- In general, Swift requires fewer lines of code than Objective C for the same task. The lines of code in both projects can be calculated by using the following command in the terminal when it is running in the same directory as the project:

```
# find . "(" -name "*.m" -or -name "*.mm" -or -name "*.h"
-or -name "*.cpp" -or -name "*.swift" ")" -print0 | xargs
-0 wc -l
```

The preceding command will search for all file formats matching `.m`, `.h`, `.swift`, and so on, recursively in all folders, and will then find the number of lines in those files, finally giving a count of the total lines of code.

The following screenshot shows the number of lines of code in Swift:

```
mare@ubuntu:~/Desktop/Projects/SecureStorage-iOSApp-main$ find . "(" -name "*.m" -or -name "*.mm" -or -name "*.h" -or -name "*.cpp" -
or -name "*.swift" ")" -print0 | xargs -0 wc -l
   33 ./SecureStorageTests/SecureStorageTests.swift
   42 ./SecureStorageUITests/SecureStorageUITests.swift
   32 ./SecureStorageUITests/SecureStorageUITestsLaunchTests.swift
  765 ./Pods/RNCryptor/Sources/RNCryptor/RNCryptor.swift
    5 ./Pods/Target Support Files/RNCryptor/RNCryptor-dummy.m
   16 ./Pods/Target Support Files/RNCryptor/RNCryptor-umbrella.h
   16 ./Pods/Target Support Files/Pods-SecureStorage/Pods-SecureStorage-umbrella.h
    5 ./Pods/Target Support Files/Pods-SecureStorage/Pods-SecureStorage-dummy.m
   36 ./SecureStorage/StoryboardExtension.swift
   81 ./SecureStorage/AppDelegate.swift
  211 ./SecureStorage/ProUserPaymentViewController.swift
   24 ./SecureStorage/ViewController.swift
  130 ./SecureStorage/Utility.swift
  183 ./SecureStorage/CardsViewController.swift
   62 ./SecureStorage/AfterLoginViewController.swift
  189 ./SecureStorage/LoginViewController.swift
   44 ./SecureStorage/Extension.swift
  118 ./SecureStorage/StringExtension.swift
   23 ./SecureStorage/DateExtension.swift
  110 ./SecureStorage/UsersViewController.swift
   25 ./SecureStorage/ViewControllerExtension.swift
   55 ./SecureStorage/SceneDelegate.swift
   36 ./SecureStorage/UIAlertControllerExtension.swift
  106 ./SecureStorage/SignUpViewController.swift
 2187 total
```

Figure 5.3 – Swift lines of code

In comparison to Swift, the following screenshot shows the lines of code for the same functionality when written in Objective C:

```
mare@ubuntu:~/Desktop/Projects/SecureStorageObjC$ find . "(" -name "*.m" -or -name "*.mm" -or -name "*.h" -or -name "*.cpp
" -or -name "*.swift" ")" -print0 | xargs -0 wc -l
   15 ./SecureStorageObjC/SceneDelegate.h
   19 ./SecureStorageObjC/Extension/NSString+Extension.m
   21 ./SecureStorageObjC/Extension/UIAlertController+Extension.h
   38 ./SecureStorageObjC/Extension/UIStoryboard+Extension.m
   42 ./SecureStorageObjC/Extension/UIAlertController+Extension.m
   18 ./SecureStorageObjC/Extension/UIView+UIView_Extension.h
   30 ./SecureStorageObjC/Extension/UIStoryboard+Extension.h
   20 ./SecureStorageObjC/Extension/NSObject+Extension.h
   19 ./SecureStorageObjC/Extension/UIViewController+Extension.h
   21 ./SecureStorageObjC/Extension/NSObject+Extension.m
   18 ./SecureStorageObjC/Extension/UIViewController+Extension.m
   19 ./SecureStorageObjC/Extension/NSString+Extension.h
   17 ./SecureStorageObjC/Extension/UIView+UIView_Extension.m
   24 ./SecureStorageObjC/Models/Card.h
   16 ./SecureStorageObjC/Models/Mappable.h
   37 ./SecureStorageObjC/Models/Card.m
   21 ./SecureStorageObjC/GeneralUtils.h
   57 ./SecureStorageObjC/SceneDelegate.m
  535 ./SecureStorageObjC/Encryption/NSData+CommonCrypto.m
  122 ./SecureStorageObjC/Encryption/FWEncryptorAES.m
  330 ./SecureStorageObjC/Encryption/NSData+Base64.m
   44 ./SecureStorageObjC/Encryption/NSData+Base64.h
   31 ./SecureStorageObjC/Encryption/FWEncryptorAES.h
  111 ./SecureStorageObjC/Encryption/NSData+CommonCrypto.h
   14 ./SecureStorageObjC/AppDelegate.h
   18 ./SecureStorageObjC/main.m
   40 ./SecureStorageObjC/AppDelegate.m
   42 ./SecureStorageObjC/ViewControllers/CardTableViewCell.h
  177 ./SecureStorageObjC/ViewControllers/ProUserPaymentViewController.m
   32 ./SecureStorageObjC/ViewControllers/AfterLoginViewController.h
  126 ./SecureStorageObjC/ViewControllers/SignUpViewController.m
   22 ./SecureStorageObjC/ViewControllers/LoginViewController.h
   21 ./SecureStorageObjC/ViewControllers/CardsViewController.h
   53 ./SecureStorageObjC/ViewControllers/CardsViewController.m
   25 ./SecureStorageObjC/ViewControllers/SignUpViewController.h
   59 ./SecureStorageObjC/ViewControllers/AfterLoginViewController.m
   28 ./SecureStorageObjC/ViewControllers/ProUserPaymentViewController.h
  112 ./SecureStorageObjC/ViewControllers/LoginViewController.m
   47 ./SecureStorageObjC/ViewControllers/CardTableViewCell.m
  119 ./SecureStorageObjC/GeneralUtils.m
 2560 total
```

Figure 5.4 – Objective C lines of code

Apart from the visible differences in the projects, there are some other differences too, such as the following:

- Swift only has classes, whereas Objective C has both classes and structs.
- Swift is comparatively faster in compilation than Objective C.
- Swift is an Apache-licensed open source project.
- Dynamic libraries are supported in Swift but not in Objective C.

In order to further understand how the two languages differ from a reverse engineering point of view, we will analyze the code snippet, which does the same task, in both languages.

## The difference between Objective C and Swift from a reverse engineering perspective

We just looked at some of the major differences between Objective C and Swift from a programming perspective. Let's also analyze the difference from a reverse engineering point of view. The following code snippets are from mobile apps doing the same task, using Objective C and Swift:

### Swift code

```swift
func loginWithUserName(userName : String, password : String){
    let str1 = userName
    let str2 = password

    UserDefaults.standard.set(str1, forKey: "userNameForLogin")
    UserDefaults.standard.set(str2, forKey: "passwordForLogin")

    let alert = UIAlertController(title: "Login here", message:
"Login Successful",            preferredStyle: .alert)

    alert.addAction(UIAlertAction(title: "Ok", style: .default,
handler: { _ in

    }))
    self.present(alert, animated: true, completion: nil)
}
```

## Objective C code

```
-(void)loginInAppUsing :(NSString*) userName and : (NSString*)
password{

    NSString* str1 = userName;
    NSString* str2 = password;

    [[NSUserDefaults standardUserDefaults] setObject:str1
forKey:@"userNameForLogin"];
    [[NSUserDefaults standardUserDefaults] setObject:str2
forKey:@"passwordForLogin"];

    UIAlertController *alert=[ UIAlertController
alertControllerWithTitle:@"Login here" message:@"Login
Successful" preferredStyle:UIAlertControllerStyleActionSheet];
    UIAlertAction *cancelaction=[UIAlertAction
actionWithTitle:@"Ok" style:UIAlertActionStyleDestructive
handler:nil];
    [alert addAction:cancelaction];
    [self presentViewController:alert animated:YES
completion:nil];

}
```

When the binaries storing the preceding code are disassembled using a tool such as Ghidra, we get the following results.

## Disassembled Swift code (entry function):

```
void entry(undefined8 param_1,undefined8 param_2)

{
  int iVar1;
  undefined8 uVar2;
  undefined8 uVar3;
  undefined8 uVar4;
  undefined8 uVar5;

  uVar2 = FUN_100005e78(0);
  uVar3 = __stubs::_$ss11CommandLineO10unsafeArgvSpySpys4Int8VGSgGvgZ();
  __stubs::_$ss11CommandLineO4argcs5Int32VvgZ();
  uVar4 = __stubs::_$ss11CommandLineO4argcs5Int32VvgZ();
  __stubs::_swift_getObjCClassFromMetadata(uVar2);
  __stubs::_NSStringFromClass();
  uVar2 = __stubs::_objc_retainAutoreleasedReturnValue();
  uVar5 = __stubs::_$sSS10FoundationE36_unconditionallyBridgeFromObjectiveCySSSo8NSStringCSgFZ();
  iVar1 = __stubs::_$s5UIKit17UIApplicationMainys5Int32VAD_SpySpys4Int8VGGSgSSSgAJtF
                   (uVar4,uVar3,0,0,uVar5,param_2);
  __stubs::_objc_release(uVar2);
  __stubs::_swift_bridgeObjectRelease(param_2);
                  /* WARNING: Subroutine does not return */
  __stubs::_exit(iVar1);
}
```

Figure 5.5 – The disassembled code of the binary with Swift code

## Disassembled Objective C code (entry function):

```
[0x100006/a8]> afl
0x1000067a8  1 136   entry0
0x100009df8  1 28    sym.func.100009df8
0x100014ffc  1 12    sym.imp.objc_opt_self
0x100014888  1 12    sym.imp._ss11CommandLineO10unsafeArgvSpySpys4Int8VGSgGvgZ
0x100014894  1 12    sym.imp._ss11CommandLineO4argcs5Int32VvgZ
0x100014aec  1 12    sym.imp.swift_getObjCClassFromMetadata
0x10001496c  1 12    sym.imp.NSStringFromClass
0x100014a2c  1 12    sym.imp.objc_retainAutoreleasedReturnValue
0x100014714  1 12    sym.imp._sSS10FoundationE36_unconditionallyBridgeFromObjectiveCySSSo8NSStringCSgFZ
0x100014684  1 12    sym.imp.UIKit.UIApplicationMain
0x100014a08  1 12    sym.imp.objc_release
0x100014a74  1 12    sym.imp.swift_bridgeObjectRelease
0x10001499c  1 12    sym.imp.exit
0x100014528  1 12    fcn.100014528
0x100014534  1 12    sym.imp.Foundation.CharacterSet
0x100014540  1 12    fcn.100014540
0x10001454c  1 12    fcn.10001454c
0x100014558  1 12    sym.imp.Foundation.Notification
0x100014564  1 12    fcn.100014564
0x100014570  1 12    sym.imp.Foundation.__DataStorage.bytes.allocator
0x10001457c  1 12    sym.imp.Foundation.__DataStorage._bytes.allocator
0x100014588  1 12    sym.imp.Foundation.__DataStorage.length.allocator
0x100014594  1 12    sym.imp.Foundation.__DataStorage._length.allocator__Swift.Int
0x1000145a0  1 12    sym.imp.Foundation.__DataStorage._offset.allocator__Swift.Int
0x1000145ac  1 12    sym.imp.Foundation.__DataStorage.allocator
0x1000145b8  1 12    sym.imp.Foundation.ContiguousBytes.withUnsafe.allocator
0x1000145c4  1 12    sym.imp.Foundation._convertErrorToNSError
0x1000145d0  1 12    sym.imp.Foundation._convertNSErrorToError
0x1000145dc  1 12    fcn.1000145dc
0x1000145e8  1 12    fcn.1000145e8
0x1000145f4  1 12    fcn.1000145f4
0x100014600  1 12    fcn.100014600
0x10001460c  1 12    fcn.10001460c
0x100014618  1 12    fcn.100014618
0x100014624  1 12    sym.imp.Foundation.Data
0x100014630  1 12    fcn.100014630
0x10001463c  1 12    fcn.10001463c
0x100014648  1 12    sym.imp.Foundation.Date
0x100014654  1 12    fcn.100014654
0x100014660  1 12    fcn.100014660
```

Figure 5.6 – The disassembled code of the binary with Objective C code

From the preceding decompiled code, we can see that the Swift version of the code is comparatively more difficult to read or analyze than the Objective C version. For simpler tasks, such as extracting class information and finding strings in the binary, it is possible to use the same tools that we use for Objective C. However, the same tools might not give the desired output a lot of the time with Swift binaries.

In the next section of this chapter, we will use an advanced reverse engineering framework known as Radare2, or r2. You can find the official GitHub repository of Radare2 here: `https://github.com/radareorg/radare2`. The Radare2 framework contains multiple tools/utilities that can be used to perform different tasks such as extracting information from executable binaries, debugging apps, assembling, and disassembling.

In the next section, we will install the Radare2 framework on our Ubuntu virtual machine and look into its use for reverse engineering a Swift-based application.

# Reverse engineering a Swift application

Before we get into reverse engineering the Swift application, here's just a reminder that in order to extract the unencrypted application from an iOS device, we will need tools such as `frida-ios-dump` (`https://github.com/AloneMonkey/frida-ios-dump`) and `frida-ipa-dump` (`https://github.com/integrity-sa/frida-ipa-dump`). You can read more about how to use them to extract encrypted applications from iOS devices in their respective GitHub repositories.

> **Important Note**
>
> Together with Frida-based tools such as `frida-ios-dump` and `frida-ipa-dump`, there are several other applications available for jailbroken devices that can help to extract an unencrypted application from the device. Most of these applications can be downloaded on a jailbroken device using Cydia (a third-party application installer for jailbroken iOS devices).

Once the application is downloaded, please follow the steps to extract the application binary from the package by unarchiving it (refer to the *Reverse engineering the iOS app* section of *Chapter 4, Reverse Engineering an iOS Application*).

We will create a directory named `swift` on our Ubuntu virtual machine and save the extracted Swift binary in that directory.

As mentioned before, to reverse engineer the Swift binary, we will be using the Radare2 framework.

# Installing the Radare2 framework

The official GitHub repository of Radare2 is located at `https://github.com/radareorg/radare2`. You can follow any of the given steps to install the framework. We will be installing it using the `pip` utility. This is because the version of the tool available in the `apt` (or other package manager such as `yum`) package might not be the latest:

1.  To install the Radare2 framework using `pip`, we first need to install the `pip` utility by running the following command:

    ```
    # sudo apt install python3-pip
    ```

    The following screenshot shows the result of the preceding command:

```
mare@ubuntu:~/Desktop/swift$ sudo apt install python3-pip
Reading package lists... Done
Building dependency tree
Reading state information... Done
The following packages were automatically installed and are no longer required:
  linux-headers-5.11.0-27-generic linux-headers-5.11.0-40-generic linux-hwe-5.11-headers-5.11.0-27
  linux-hwe-5.11-headers-5.11.0-40 linux-image-5.11.0-27-generic linux-image-5.11.0-40-generic
  linux-modules-5.11.0-27-generic linux-modules-5.11.0-40-generic linux-modules-extra-5.11.0-27-generic
  linux-modules-extra-5.11.0-40-generic
Use 'sudo apt autoremove' to remove them.
The following NEW packages will be installed:
  python3-pip
0 upgraded, 1 newly installed, 0 to remove and 91 not upgraded.
Need to get 231 kB of archives.
After this operation, 1,050 kB of additional disk space will be used.
Get:1 http://us.archive.ubuntu.com/ubuntu focal-updates/universe amd64 python3-pip all 20.0.2-5ubuntu1.6 [231
kB]
Fetched 231 kB in 3s (72.1 kB/s)
Selecting previously unselected package python3-pip.
(Reading database ... 269629 files and directories currently installed.)
Preparing to unpack .../python3-pip_20.0.2-5ubuntu1.6_all.deb ...
Unpacking python3-pip (20.0.2-5ubuntu1.6) ...
Setting up python3-pip (20.0.2-5ubuntu1.6) ...
Processing triggers for man-db (2.9.1-1) ...
```

Figure 5.7 – Installing pip

2.  Once pip has been successfully installed, run the following code to install Radare2:

    ```
    # pip install r2env
    ```

    Here's the output:

```
mare@ubuntu:~/Desktop/swift$ pip install r2env
Processing /home/mare/.cache/pip/wheels/c7/43/fb/ba404e85b3ae3b6cf50f2b5fdd461797724db44b7744680faf/r2env-0.4.
1-py3-none-any.whl
Requirement already satisfied: GitPython in /home/mare/.local/lib/python3.8/site-packages (from r2env) (3.1.25
)
Requirement already satisfied: dploy>=0.1.2 in /home/mare/.local/lib/python3.8/site-packages (from r2env) (0.1
.2)
Requirement already satisfied: wget in /home/mare/.local/lib/python3.8/site-packages (from r2env) (3.2)
Requirement already satisfied: colorama>=0.4.4 in /home/mare/.local/lib/python3.8/site-packages (from r2env) (
0.4.4)
Requirement already satisfied: gitdb<5,>=4.0.1 in /home/mare/.local/lib/python3.8/site-packages (from GitPytho
n->r2env) (4.0.9)
Requirement already satisfied: smmap<6,>=3.0.1 in /home/mare/.local/lib/python3.8/site-packages (from gitdb<5,
>=4.0.1->GitPython->r2env) (5.0.0)
Installing collected packages: r2env
  WARNING: The script r2env is installed in '/home/mare/.local/bin' which is not on PATH.
  Consider adding this directory to PATH or, if you prefer to suppress this warning, use --no-warn-script-loca
tion.
Successfully installed r2env-0.4.1
```

Figure 5.8 – Installing r2env

3.  Once installed successfully, you can verify it by running the following command:

```
# r2 -v
```

This will show the installed version of r2:

```
mare@ubuntu:~/Desktop/swift$ r2 -v
radare2 4.2.1 0 @ linux-x86-64 git.4.2.1
commit: unknown build:      _
```

Figure 5.9 – Verifying the r2 version

Once Radare2 is installed on your virtual machine, let's use the tool to reverse engineer the Swift application SecureStorage.

# Using the Radare2 framework to reverse engineer a Swift application

The Radare2 framework contains different tools for different purposes. Some of these tools are as follows:

- radare2: This is the main tool of the whole framework, and its main purpose is to perform disassembling, binary patching, data comparison, and more on several different types of input/output sources and files.

- rabin2: A tool to extract information from executable binaries, such as Mach-O.

- rasm2: An assembler and disassembler for different architectures.

You can read more about the other tools of the framework in the official documentation: https://book.rada.re/.

We will be using the tools of the Radare2 framework to extract information from the binaries as well as to find any sensitive data stored.

To start with, we can analyze the metadata of a Swift binary file. Analyzing the metadata can help in identifying the binary architecture, finding out whether the binary is stripped, or an ELF type, whether there are binary protections, and more. Once you download the SecureStorage IPA file (`https://github.com/0ctac0der/SecureStorage-iOSApp/releases/download/0.1/SecureStorage.ipa`), extract the binary from it and then run the following command to analyze the binary using r2:

```
# r2 SecureStorage
# i
```

The following screenshot shows the result of the preceding command:

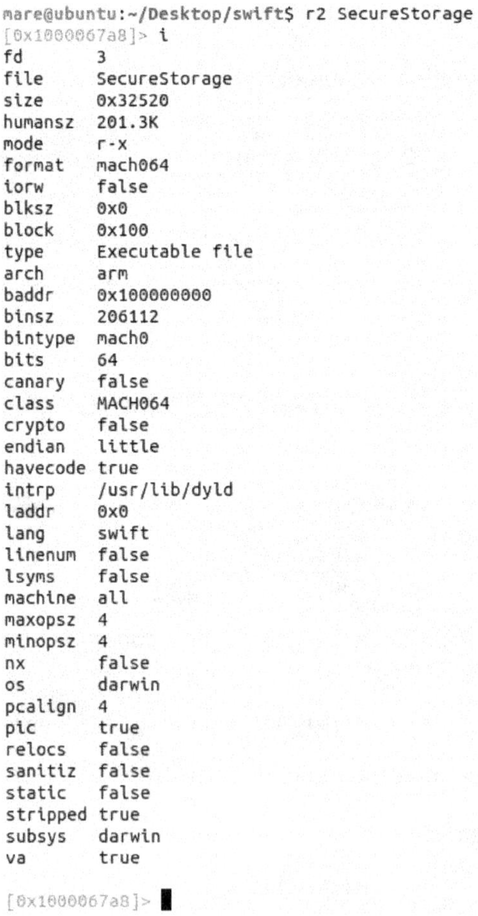

```
mare@ubuntu:~/Desktop/swift$ r2 SecureStorage
[0x1000067a8]> i
fd       3
file     SecureStorage
size     0x32520
humansz  201.3K
mode     r-x
format   mach064
iorw     false
blksz    0x0
block    0x100
type     Executable file
arch     arm
baddr    0x100000000
binsz    206112
bintype  mach0
bits     64
canary   false
class    MACH064
crypto   false
endian   little
havecode true
intrp    /usr/lib/dyld
laddr    0x0
lang     swift
linenum  false
lsyms    false
machine  all
maxopsz  4
minopsz  4
nx       false
os       darwin
pcalign  4
pic      true
relocs   false
sanitiz  false
static   false
stripped true
subsys   darwin
va       true

[0x1000067a8]> █
```

Figure 5.10 – Using r2 to extract metadata

In the preceding code snippet, with `r2 SecureStorage`, we run Radare2 on the Swift binary of the SecureStorage application, and then with the `i` command, we extract the information about the binary. The extracted information has some interesting details about the binary, such as architecture, format, and protections (NX, canary, and relocs (indicates that the binary performs runtime relocation)).

## Extracting strings from the binary using rabin2

You can also extract other interesting information, such as symbols and strings, from the binary using the different utilities in the Radare2 framework, such as `rabin2`:

```
# rabin2 -zzz SecureStorage
```

Here, `-zzz` stands for *show strings* (like a gnu "strings" utility does).

The following screenshot displays the output:

```
mare@ubuntu:~/Desktop/swift$ rabin2 -zzz SecureStorage
000 0x00000028 0x00000028  10  11 () ascii __PAGEZERO
001 0x00000070 0x00000070   6   7 () ascii __TEXT
002 0x000000b0 0x000000b0   6   7 () ascii __text
003 0x000000c0 0x000000c0   6   7 () ascii __TEXT
004 0x00000100 0x00000100   7   8 () ascii __stubs
005 0x00000110 0x00000110   6   7 () ascii __TEXT
006 0x00000150 0x00000150  15  16 () ascii __objc_methlist
007 0x00000160 0x00000160   6   7 () ascii __TEXT
008 0x000001a0 0x000001a0   7   8 () ascii __const
009 0x000001b0 0x000001b0   6   7 () ascii __TEXT
010 0x000001f0 0x000001f0  14  15 () ascii __swift5_entry
011 0x00000200 0x00000200   6   7 () ascii __TEXT
012 0x00000240 0x00000240   9  10 () ascii __cstring
013 0x00000250 0x00000250   6   7 () ascii __TEXT
014 0x00000290 0x00000290  15  16 () ascii __objc_methname
015 0x000002a0 0x000002a0   6   7 () ascii __TEXT
016 0x000002e0 0x000002e0  22  23 () ascii __swift5_typeref__TEXT
017 0x00000330 0x00000330  22  23 () ascii __swift5_reflstr__TEXT
018 0x00000380 0x00000380  22  23 () ascii __swift5_fieldmd__TEXT
019 0x000003d0 0x000003d0  14  15 () ascii __swift5_types
020 0x000003e0 0x000003e0   6   7 () ascii __TEXT
021 0x00000420 0x00000420  22  23 () ascii __swift5_capture__TEXT
022 0x00000470 0x00000470  15  16 () ascii __swift5_protos
023 0x00000480 0x00000480   6   7 () ascii __TEXT
024 0x000004c0 0x000004c0  14  15 () ascii __swift5_proto
025 0x000004d0 0x000004d0   6   7 () ascii __TEXT
026 0x00000510 0x00000510  13  14 () ascii __unwind_info
027 0x00000520 0x00000520   6   7 () ascii __TEXT
028 0x00000560 0x00000560  10  11 () ascii __eh_frame
029 0x00000570 0x00000570   6   7 () ascii __TEXT
030 0x000005b8 0x000005b8  12  13 () ascii __DATA_CONST
```

Figure 5.11 – Using rabin2 to extract strings

## Extracting strings from the binary using r2

Next, let's look at the `r2` utility:

```
# iz
```

Here's the result:

```
[0x1000067a8]> iz
[Strings]
nth paddr       vaddr      len size section          type    string
―――――――――――――――――――――――――――――――――――――――――――――――――――――――――――――――――――――――――――
0   0x000153d0 0x1000153d0 13  14   3.__TEXT.__const  ascii   SecureStorage
1   0x000153f0 0x1000153f0 20  21   3.__TEXT.__const  ascii   SignUpViewController
2   0x000154d0 0x1000154d0 24  25   3.__TEXT.__const  ascii   AfterLoginViewController
3   0x000155b0 0x1000155b0 14  15   3.__TEXT.__const  ascii   ViewController
4   0x000155f0 0x1000155f0 13  14   3.__TEXT.__const  ascii   SecureStorage
5   0x00015600 0x100015600 23  24   3.__TEXT.__const  ascii   CardsDetailCellDelegate
6   0x0001564c 0x10001564c 15  16   3.__TEXT.__const  ascii   CardsDetailCell
7   0x00015678 0x100015678 4   20   3.__TEXT.__const  utf32le 1'\t\n
8   0x00015780 0x100015780 19  20   3.__TEXT.__const  ascii   CardsViewController
9   0x00015820 0x100015820 11  12   3.__TEXT.__const  ascii   AppDelegate
10  0x000158a0 0x1000158a0 13  14   3.__TEXT.__const  ascii   SecureStorage
11  0x000158b0 0x1000158b0 22  23   3.__TEXT.__const  ascii   UserDetailCellDelegate
12  0x000158f0 0x1000158f0 4   5    3.__TEXT.__const  ascii   User
13  0x00015934 0x100015934 14  15   3.__TEXT.__const  ascii   UserDetailCell
14  0x00015960 0x100015960 4   20   3.__TEXT.__const  utf32le /%\t\n
15  0x00015a60 0x100015a60 19  20   3.__TEXT.__const  ascii   UsersViewController
16  0x00015a90 0x100015a90 4   20   3.__TEXT.__const  utf32le * \b\n
17  0x00015b80 0x100015b80 13  14   3.__TEXT.__const  ascii   SecureStorage
18  0x00015b90 0x100015b90 28  29   3.__TEXT.__const  ascii   ProUserPaymentViewController
19  0x00015bcc 0x100015bcc 4   20   3.__TEXT.__const  utf32le .$\b\n
20  0x00015ce0 0x100015ce0 13  14   3.__TEXT.__const  ascii   SecureStorage
21  0x00015cf0 0x100015cf0 19  20   3.__TEXT.__const  ascii   LoginViewController
22  0x00015dc0 0x100015dc0 13  14   3.__TEXT.__const  ascii   SceneDelegate
23  0x00015e60 0x100015e60 7   8    3.__TEXT.__const  ascii   Utility
0   0x00015f20 0x100015f20 41  42   5.__TEXT.__cstring ascii  _TtC13SecureStorage20SignUpViewController
1   0x00015f50 0x100015f50 13  14   5.__TEXT.__cstring ascii  @32@0:8@16@24
2   0x00015f69 0x100015f69 13  14   5.__TEXT.__cstring ascii  selectedIndex
3   0x00015f78 0x100015f78 12  13   5.__TEXT.__cstring ascii  txtfUserName
4   0x00015f85 0x100015f85 12  13   5.__TEXT.__cstring ascii  txtfPassword
5   0x00015fa0 0x100015fa0 19  20   5.__TEXT.__cstring ascii  txtfConfirmPassword
6   0x00015fb4 0x100015fb4 9   10   5.__TEXT.__cstring ascii  btnSignUp
7   0x00015fc0 0x100015fc0 33  34   5.__TEXT.__cstring ascii  T@"UITextField",N,W,VtxtfUserName
8   0x00015ff0 0x100015ff0 33  34   5.__TEXT.__cstring ascii  T@"UITextField",N,W,VtxtfPassword
9   0x00016020 0x100016020 40  41   5.__TEXT.__cstring ascii  T@"UITextField",N,W,VtxtfConfirmPassword
10  0x00016050 0x100016050 27  28   5.__TEXT.__cstring ascii  T@"UIButton",N,W,VbtnSignUp
11  0x0001606c 0x10001606c 14  15   5.__TEXT.__cstring ascii  SecureStorage1
12  0x00016080 0x100016080 23  24   5.__TEXT.__cstring ascii  Please Enter all values
13  0x000160a0 0x1000160a0 21  22   5.__TEXT.__cstring ascii  Password not matching
14  0x000160c0 0x1000160c0 16  17   5.__TEXT.__cstring ascii  UserNameForLogin
```

Figure 5.12 – Using r2 to extract strings

## Extracting symbols from the binary

Next, let's learn how to extract symbols:

```
# is
```

The following screenshot shows the result of the preceding command:

```
[0x1000067a8]> is
[Symbols]

nth paddr       vaddr       bind   type size lib name
------------------------------------------------------------------------------------
0    0x00000000 0x100000000 GLOBAL FUNC 0        __mh_execute_header
1    0x05614542 0x05614542  LOCAL  FUNC 0        radr://5614542
2    0x00014528 0x100014528 LOCAL  FUNC 0        Foundation.CharacterSet
3    0x00014534 0x100014534 LOCAL  FUNC 0        Foundation.CharacterSet
4    0x00014540 0x100014540 LOCAL  FUNC 0        Foundation.Notification
5    0x0001454c 0x10001454c LOCAL  FUNC 0        Foundation.Notification
6    0x00014558 0x100014558 LOCAL  FUNC 0        Foundation.Notification
7    0x00014564 0x100014564 LOCAL  FUNC 0        Foundation.__DataStorage.bytes.allocator
8    0x00014570 0x100014570 LOCAL  FUNC 0        Foundation.__DataStorage.bytes.allocator
9    0x0001457c 0x10001457c LOCAL  FUNC 0        Foundation.__DataStorage._bytes.allocator -> ()
10   0x00014588 0x100014588 LOCAL  FUNC 0        Foundation.__DataStorage.length.allocator
11   0x00014594 0x100014594 LOCAL  FUNC 0        Foundation.__DataStorage._length.allocator__Swift.Int -> ()
12   0x000145a0 0x1000145a0 LOCAL  FUNC 0        Foundation.__DataStorage._offset.allocator__Swift.Int -> ()
13   0x000145ac 0x1000145ac LOCAL  FUNC 0        Foundation.__DataStorage.allocator
14   0x000145b8 0x1000145b8 LOCAL  FUNC 0        Foundation.ContiguousBytes.withUnsafe.allocator
15   0x000145c4 0x1000145c4 LOCAL  FUNC 0        Foundation._convertErrorToNSError
16   0x000145d0 0x1000145d0 LOCAL  FUNC 0        Foundation._convertNSErrorToError
17   0x000145dc 0x1000145dc LOCAL  FUNC 0        Foundation.Data
18   0x000145e8 0x1000145e8 LOCAL  FUNC 0        Foundation.Data
19   0x000145f4 0x1000145f4 LOCAL  FUNC 0        Foundation.Data
20   0x00014600 0x100014600 LOCAL  FUNC 0        Foundation.Data
21   0x0001460c 0x10001460c LOCAL  FUNC 0        Foundation.Data
22   0x00014618 0x100014618 LOCAL  FUNC 0        Foundation.Data
23   0x00014624 0x100014624 LOCAL  FUNC 0        Foundation.Data
24   0x00014630 0x100014630 LOCAL  FUNC 0        Foundation.Date
25   0x0001463c 0x10001463c LOCAL  FUNC 0        Foundation.Date
26   0x00014648 0x100014648 LOCAL  FUNC 0        Foundation.Date
27   0x00014654 0x100014654 LOCAL  FUNC 0        Foundation.IndexPath
28   0x00014660 0x100014660 LOCAL  FUNC 0        Foundation.IndexPath
29   0x0001466c 0x10001466c LOCAL  FUNC 0        Foundation.IndexPath
30   0x00014678 0x100014678 LOCAL  FUNC 0        Foundation.IndexPath
31   0x00014684 0x100014684 LOCAL  FUNC 0        UIKit.UIApplicationMain
32   0x00014690 0x100014690 LOCAL  FUNC 0        Dispatch
33   0x0001469c 0x10001469c LOCAL  FUNC 0        Dispatch
34   0x000146a8 0x1000146a8 LOCAL  FUNC 0        Dispatch
35   0x000146b4 0x1000146b4 LOCAL  FUNC 0        Dispatch
36   0x000146c0 0x1000146c0 LOCAL  FUNC 0        Dispatch
37   0x000146cc 0x1000146cc LOCAL  FUNC 0        Dispatch.p
```

Figure 5.13 – Using r2 to extract symbols

Similarly, we can also extract all the functions being used in the binary by using the
`afl` command. But before doing that, we need to run the `aaa` command to analyze all
referenced code:

```
[0x1000067a8]> aaa
[x] Analyze all flags starting with sym. and entry0 (aa)
[x] Analyze function calls (aac)
[x] Analyze len bytes of instructions for references (aar)
[x] Check for objc references
[x] Parsing metadata in ObjC to find hidden xrefs
[x] A total of 0 xref were found
[x] Set 185 dwords at 0x100021b80
[x] Check for vtables
[x] Finding xrefs in noncode section with anal.in=io.maps
[x] Analyze value pointers (aav)
[x] Value from 0x100000000 to 0x10001c000 (aav)
[x] 0x100000000-0x10001c000 in 0x100000000-0x10001c000 (aav)
[x] Emulate code to find computed references (aae)
[x] Type matching analysis for all functions (aaft)
[x] Propagate noreturn information
[x] Use -AA or aaaa to perform additional experimental analysis.
```

Figure 5.14 – Analyzing the binary file

Once the binary file has been analyzed, we can then run the `afl` command to extract all the functions. Refer to the following screenshot for the same:

```
[0x1000067a8]> afl
0x1000067a8   1 136        entry0
0x100009df8   1 28         sym.func.100009df8
0x1000149fc   1 12         sym.imp.objc_opt_self
0x100014888   1 12         sym.imp._ss11CommandLineO10unsafeArgvSpySpys4Int8VGSgGvgZ
0x100014894   1 12         sym.imp._ss11CommandLineO04argcs5Int32VvgZ
0x100014aec   1 12         sym.imp.swift_getObjCClassFromMetadata
0x10001496c   1 12         sym.imp.NSStringFromClass
0x100014a2c   1 12         sym.imp.objc_retainAutoreleasedReturnValue
0x100014714   1 12         sym.imp._sSS10FoundationE36_unconditionallyBridgeFromObjectiveCySSSo8NSStringCSgFZ
0x100014684   1 12         sym.imp.UIKit.UIApplicationMain
0x100014a08   1 12         sym.imp.objc_release
0x100014a74   1 12         sym.imp.swift_bridgeObjectRelease
0x10001499c   1 12         sym.imp.exit
0x100014528   1 12         fcn.100014528
0x100014534   1 12         sym.imp.Foundation.CharacterSet
0x100014540   1 12         fcn.100014540
0x10001454c   1 12         fcn.10001454c
0x100014558   1 12         sym.imp.Foundation.Notification
0x100014564   1 12         fcn.100014564
0x100014570   1 12         sym.imp.Foundation.__DataStorage.bytes.allocator
0x10001457c   1 12         sym.imp.Foundation.__DataStorage._bytes.allocator
0x100014588   1 12         sym.imp.Foundation.__DataStorage.length.allocator
0x100014594   1 12         sym.imp.Foundation.__DataStorage._length.allocator__Swift.Int
0x1000145a0   1 12         sym.imp.Foundation.__DataStorage._offset.allocator__Swift.Int
0x1000145ac   1 12         sym.imp.Foundation.__DataStorage.allocator
0x1000145b8   1 12         sym.imp.Foundation.ContiguousBytes.withUnsafe.allocator
0x1000145c4   1 12         sym.imp.Foundation._convertErrorToNSError
0x1000145d0   1 12         sym.imp.Foundation._convertNSErrorToError
0x1000145dc   1 12         fcn.1000145dc
0x1000145e8   1 12         fcn.1000145e8
0x1000145f4   1 12         fcn.1000145f4
0x100014600   1 12         fcn.100014600
0x10001460c   1 12         fcn.10001460c
0x100014618   1 12         fcn.100014618
0x100014624   1 12         sym.imp.Foundation.Data
0x100014630   1 12         fcn.100014630
0x10001463c   1 12         fcn.10001463c
0x100014648   1 12         sym.imp.Foundation.Date
0x100014654   1 12         fcn.100014654
0x100014660   1 12         fcn.100014660
```

Figure 5.15 – Finding all the functions in the binary

Once you have some details about the binary, the next step is to find the entry point and the `main` function. Both of these can be found by the `ie` and `iM` commands, respectively (refer to the following screenshot):

```
[0x1000067a8]> ie
[Entrypoints]
vaddr=0x1000067a8 paddr=0x000067a8 haddr=0x00000b40 type=program

1 entrypoints

[0x1000067a8]> iM
[Main]
vaddr=0x1000067a8 paddr=0x1000067a8
```

Figure 5.16 – Finding the entry point and the main address

Let's say we are interested in finding where a specific string in the binary is being referenced. One of the extracted strings is a very long string.

```
5.__TEXT.__cstring         ascii   v32@0:8@"UIScene"16@"NSUserActivity"24
5.__TEXT.__cstring         ascii   v32@0:8@"UIScene"16@"NSString"24
5.__TEXT.__cstring         ascii   v40@0:8@"UIScene"16@"NSString"24@"NSError"32
5.__TEXT.__cstring         ascii    TtC13SecureStorage7Utility
5.__TEXT.__cstring         ascii   8523GF8WGFE3272357538782F413F4428472B7826DHH9W8EF832R923R9CH892FH9
5.__TEXT.__cstring         ascii   Array convertIntoJSON -
5.__TEXT.__cstring         ascii   Failed to load:
6.__TEXT.__objc_methname ascii   txtfUserName
```

Figure 5.17 – An interesting string extracted

Let's find out where this string is present in the binary. To do so, we will run the following command:

```
# /
8873456HFEIFHFIUDFHT38256239R9F8SDCFUWEYGR87465823568EWIFH89T5
```

Here's the output:

```
[0x100000a8]> / 8873456HFEIFHFIUDFHT38256239R9F8SDCFUWEYGR87465823568EWIFH89T5
Searching 62 bytes in [0x100028000-0x100034000]
hits: 0
Searching 62 bytes in [0x100020000-0x100028000]
hits: 0
Searching 62 bytes in [0x10001c000-0x100020000]
hits: 0
Searching 62 bytes in [0x100000000-0x10001c000]
hits: 1
0x1000186c0 hit0_0 .age7Utility8873456HFEIFHFIUDFHT38256239R9F8SDCFUWEYGR87465823568EWIFH89T5Array convertI.
```

Figure 5.18 – Finding the string reference

The search resulted in four addresses where the string is present. The axt command can be used to find where this string is being referenced by providing the address of the string (shown in the preceding screenshots) – that is, 0x1000186c0. But before running the axt command, we need to run the aaa command (if it has not been done already) to analyze all referenced code:

```
# aaa
```

The following screenshot shows the result of the preceding command:

```
[0x1000067a8]> aaa
[x] Analyze all flags starting with sym. and entry0 (aa)
[x] Analyze function calls (aac)
[x] Analyze len bytes of instructions for references (aar)
[x] Check for objc references
[x] Parsing metadata in ObjC to find hidden xrefs
[x] A total of 0 xref were found
[x] Set 185 dwords at 0x100021b80
[x] Check for vtables
[x] Finding xrefs in noncode section with anal.in=io.maps
[x] Analyze value pointers (aav)
[x] Value from 0x100000000 to 0x10001c000 (aav)
[x] 0x100000000-0x10001c000 in 0x100000000-0x10001c000 (aav)
[x] Emulate code to find computed references (aae)
[x] Type matching analysis for all functions (aaft)
[x] Propagate noreturn information
[x] Use -AA or aaaa to perform additional experimental analysis.
```

Figure 5.19 – Analyzing all referenced code

Once the analysis is done, we can use the axt command to search for all the places where the specific string (0x1000186c0) is being referenced:

```
# axt 0x1000186c0
```

Here's the result:

```
mare@ubuntu:~/Desktop/swift$ radare2 SecureStorage
[0x1000006afc]> aaa

[x] Analyze all flags starting with sym. and entry0 (aa)
[x] Analyze function calls (aac)
[x] Analyze len bytes of instructions for references (aar)
[x] Check for objc references
[x] Parsing metadata in ObjC to find hidden xrefs
[x] A total of 0 xref were found
[x] Set 185 dwords at 0x100022b40
[x] Check for vtables
[x] Finding xrefs in noncode section with anal.in=io.maps
[x] Analyze value pointers (aav)
[x] Value from 0x100000000 to 0x10001c000 (aav)
[x] 0x100000000-0x10001c000 in 0x100000000-0x10001c000 (aav)
[x] Emulate code to find computed references (aae)
[x] Type matching analysis for all functions (aaft)
[x] Propagate noreturn information
[x] Use -AA or aaaa to perform additional experimental analysis.
[0x1000006afc]> axt 0x1000186c0

sym.func.100012a30 0x100012c54 [STRING] adr x8, str.8523GF8WGFE3272357538782F413F4428472B7826DHH9W8EF832R923R9CH892FH9
sym.func.100013358 0x10001342c [STRING] adr x8, str.8523GF8WGFE3272357538782F413F4428472B7826DHH9W8EF832R923R9CH892FH9
[0x1000006afc]> █
```

Figure 5.20 – The string references

To further reverse engineer the binary and analyze the disassembled code, we can start with disassembling the main function by running the following command(s):

```
# s main
```

```
# pdf
```

The following screenshot shows all these commands being run on the binary:

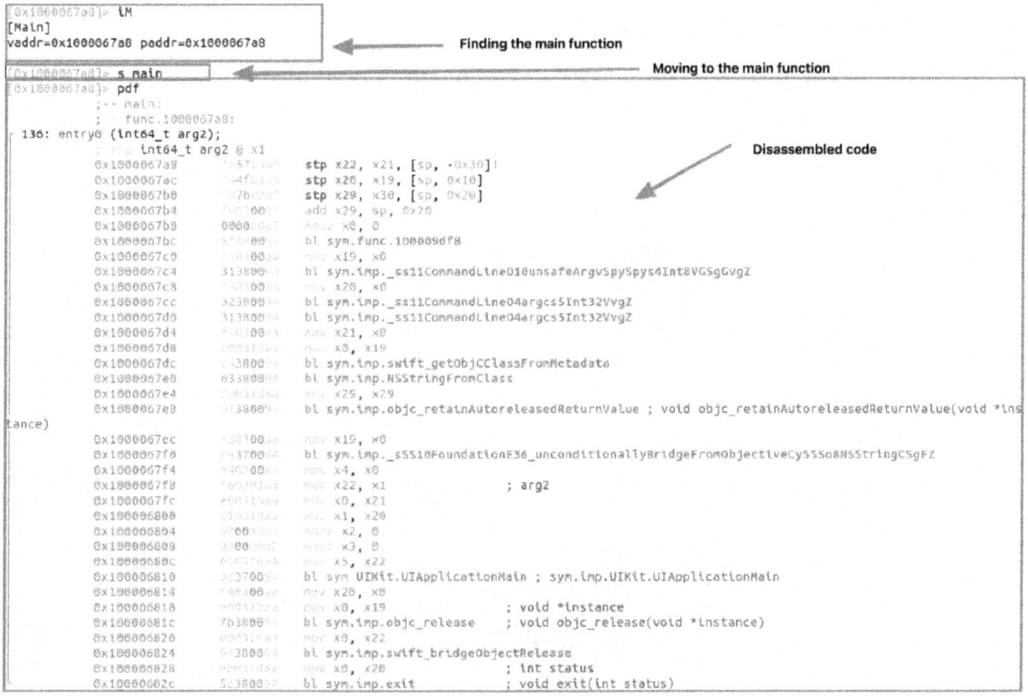

Figure 5.21 – Disassembling the main function

In order to understand the preceding disassembled code further, we can also use the `pdfs` command to show a summary of this function.

```
[0x1000067a8]> pdfs
;-- main:
;-- func.1000067a8:
0x1000067bc bl sym.func.100009df8
0x1000067c4 bl sym.imp._ss11CommandLine010unsafeArgvSpySpys4Int8VGSgGvgZ
0x1000067cc bl sym.imp._ss11CommandLine04argcs5Int32VvgZ "18" ; "P"
0x1000067d0 bl sym.imp._ss11CommandLine04argcs5Int32VvgZ
0x1000067dc bl sym.imp.swift_getObjCClassFromMetadata "c8" ; "P"
0x1000067e0 bl sym.imp.NSStringFromClass
0x1000067e8 bl sym.imp.objc_retainAutoreleasedReturnValue
0x1000067f0 bl sym.imp._sSS10FoundationE36_unconditionallyBridgeFromObjectiveCySSSo8NSStringCSgFZ
0x1000067f8 arg2
0x100006810 bl sym UIKit.UIApplicationMain
0x100006818 void *instance
0x10000681c bl sym.imp.objc_release
0x100006824 bl sym.imp.swift_bridgeObjectRelease
0x100006828 int status
0x10000682c bl sym.imp.exit
[0x1000067a8]> ▮
```

Figure 5.22 – A function summary

To analyze a function in visual mode or with a more visual presentation, we can use visual mode, using the VV command. To do so, you first need to move the pointer to the function and then run the VV command.

It is a good idea to run the afl command first and list the functions.

```
[0x100005ac0]> afl
0x1000067a8    1 136      entry0
0x100009df8    1 28       sym.func.100009df8
0x100014lfc    1 12       sym.imp.objc_opt_self
0x100014888    1 12       sym.imp._ss11CommandLine010unsafeArgvSpySpys4Int8VGSgGvgZ
0x100014894    1 12       sym.imp._ss11CommandLine04argcs5Int32VvgZ
0x100014aec    1 12       sym.imp.swift_getObjCClassFromMetadata
0x10001496c    1 12       sym.imp.NSStringFromClass
0x100014a2c    1 12       sym.imp.objc_retainAutoreleasedReturnValue
0x100014714    1 12       sym.imp._sSS10FoundationE36_unconditionallyBridgeFromObjectiveCySSSo8NSStringCSgFZ
0x100014684    1 12       sym.imp.UIKit.UIApplicationMain
0x100014a08    1 12       sym.imp.objc_release
0x100014a74    1 12       sym.imp.swift_bridgeObjectRelease
0x10001499c    1 12       sym.imp.exit
0x100014528    1 12       fcn.100014528
0x100014534    1 12       sym.imp.Foundation.CharacterSet
0x100014540    1 12       fcn.100014540
0x10001454c    1 12       fcn.10001454c
0x100014558    1 12       sym.imp.Foundation.Notification
0x100014564    1 12       fcn.100014564
0x100014570    1 12       sym.imp.Foundation.__DataStorage.bytes.allocator
0x10001457c    1 12       sym.imp.Foundation.__DataStorage._bytes.allocator
0x100014588    1 12       sym.imp.Foundation.__DataStorage.length.allocator
0x100014594    1 12       sym.imp.Foundation.__DataStorage._length.allocator__Swift.Int
0x1000145a0    1 12       sym.imp.Foundation.__DataStorage._offset.allocator__Swift.Int
0x1000145ac    1 12       sym.imp.Foundation.__DataStorage.allocator
0x1000145b8    1 12       sym.imp.Foundation.ContiguousBytes.withUnsafe.allocator
0x1000145c4    1 12       sym.imp.Foundation._convertErrorToNSError
0x1000145d0    1 12       sym.imp.Foundation._convertNSErrorToError
0x1000145dc    1 12       fcn.1000145dc
0x1000145e8    1 12       fcn.1000145e8
0x1000145f4    1 12       fcn.1000145f4
0x100014600    1 12       fcn.100014600
0x10001460c    1 12       fcn.10001460c
0x100014618    1 12       fcn.100014618
0x100014624    1 12       sym.imp.Foundation.Data
0x100014630    1 12       fcn.100014630
0x10001463c    1 12       fcn.10001463c
0x100014648    1 12       sym.imp.Foundation.Date
0x100014654    1 12       fcn.100014654
0x100014660    1 12       fcn.100014660
0x10001466c    1 12       fcn.10001466c
0x100014678    1 12       sym.imp.Foundation.IndexPath
0x100014690    1 12       fcn.100014690
0x10001469c    1 12       fcn.10001469c
0x1000146a8    1 12       fcn.1000146a8
```

Figure 5.23 – Extracting functions

Once we have the list of functions, we can choose a function and then move the pointer to that specific function:

\# s sym.func.100005ac0 (here, we are moving the pointer to the function)

\# VV (starting visual mode for that specific function)

Here's the output:

```
[0x100005ac0]> 0x100005ac0 # sym.func.100005ac0 (int64_t arg1, int64_t arg3, int64_t arg4);
```

```
[0x100005ac0]
204: sym.func.100005ac0 (int64_t arg1, int64_t arg3, int64_t arg4);
; arg int64_t arg1 @ x0
; arg int64_t arg3 @ x2
; arg int64_t arg4 @ x3
; d3
adr x3, 0x100024d8
nop
b sym.func.100005b50
```

```
0x100005b50 [od]
; CODE XREF from sym.func.100005ac0 @ 0x100005ac8
; CODE XREF from sym.func.100005aec @ 0x100005af4
; CODE XREF from sym.func.100005b18 @ 0x100005b20
; CODE XREF from sym.func.100005b44 @ 0x100005b4c
;-- func.100005b50:
stp x22, x21, [sp, -0x30]!
stp x20, x19, [sp, 0x10]
stp x29, x30, [sp, 0x20]
add x29, sp, 0x20
; arg3
mov x21, x2
; arg1
mov x20, x0
; arg4
ldr x8, [x3]
add x0, x0, x8
; arg3
mov x1, x2
bl sym.imp.swift_unknownObjectWeakAssign;[ob]
bl sym.imp.swift_unknownObjectWeakLoadStrong;[oc]
cbz x0, 0x100005c0c
```

Figure 5.24 – Analyzing the disassembled function in visual mode

We can scroll further down to see the flow of logic and understand the function better.

```
[0x100005ac0]> 0x100005ac0 # sym.func.100005ac0 (int64_t arg1, int64_t arg3, int64_t arg4);
```

```
0x100005b80 [oi]
mov x19, x0
nop
ldr x22, 0x100021f98
; void *instance
mov x0, x21
; void objc_retain(void *instance)
bl sym.imp.objc_retain;[oe]
mov x21, x0
; void *instance
mov x0, x20
; void objc_retain(void *instance)
bl sym.imp.objc_retain;[oe]
mov x20, x0
; void *instance
mov x0, x19
; char *selector
mov x1, x22
; void *objc_msgSend(void *instance, char *selector)
bl sym.imp.objc_msgSend;[of]
mov x29, x29
; void objc_retainAutoreleasedReturnValue(void *instance)
bl sym.imp.objc_retainAutoreleasedReturnValue;[og]
mov x22, x0
nop
; char *selector
ldr x1, 0x100021fa8
fmov d0, 4.00000000
; void *objc_msgSend(void *instance, char *selector)
bl sym.imp.objc_msgSend;[of]
; void *instance
mov x0, x22
; void objc_release(void *instance)
bl sym.imp.objc_release;[oh]
nop
; char *selector
ldr x1, 0x100021fa8
; void *instance
mov x0, x19
movz w2, 0x1
```

```
0x100005c0c [oj]
; CODE XREF from sym.func.100005ac0 @ 0x100005b7c
brk 0x1
```

Figure 5.25 – Analyzing the disassembled function in visual mode

The preceding steps can be used to further disassemble different functions and methods in the application binary and then look for interesting parts. Radare2 has numerous options to allow us to play more with functions, methods, classes, and so on. Some interesting things to check, from a security perspective, could be the following:

- Finding whether the binary has position-independent code: # i~pic.

> **Important Note**
>
> Position-independent code is the machine code that can execute properly (if in the primary memory) regardless of its absolute address and can be executed at any memory address without modification.

- Checking for a non-executable stack: # i~nx

> **Important Note**
>
> **Non-Executable Stack (NX)** is virtual memory protection implemented to block code injection by restricting a particular memory and implementing the NX bit.

- Checking for a stack canary: # i~canary

> **Important Note**
>
> A stack canary is a defense against memory corruption attacks. It is a value on the stack that will be overwritten by a stack buffer that overflows to the return address. When the function returns, the integrity of the canary can be verified to detect any overflow.

*Figure 5.26* shows the result of executing the previous commands:

```
[0x100006afc]> i~pic
pic         true
[0x100006afc]> i~nx
nx          false
[0x100006afc]> i~canary
canary      false _
```

Figure 5.26 – Finding interesting bits

Radare2 can also be used for more advanced tasks during a penetration test, such as binary patching and dynamic analysis.

# Summary

This chapter explained Swift programming and Swift-based iOS applications (binaries). To reverse engineer a Swift-based application binary, we can use previously used tools such as Hopper or Ghidra. But in this chapter, we saw how to use the Radare2 framework and explore the internals of the binary, extract useful information, and navigate to the code logic/flow.

As we saw that the iOS applications have compiled binaries, which are used during reverse engineering, to analyze the binaries further, we would need the knowledge of binary reverse engineering as well as how the processor instructions work. Generally, the reverse engineering of mobile applications is done during a penetration test, malware analysis, and so on. Understanding the assembly instructions and different architecture is very much a prerequisite for advanced reverse engineering. The more you explore these fantastic tools and perform reverse engineering of different types of applications, the more you will end up learning. Different applications are built with different logic and program flow; that's why there is never a set of steps or a checklist to follow while analyzing applications for security issues. However, a lot of answers can be discovered by diving deeper into the analysis through reverse engineering.

In the next chapter, we will look into the open source and commercially available tools for reverse engineering (specifically for mobile applications). We will also look into the details of the features that an open source reverse engineering framework might lack.

# Section 3: Automating Some Parts of the Reverse Engineering Process

This section talks about different open source and commercial reverse engineering tools, ways to automate some parts of reverse engineering, and vulnerability discovery in mobile apps. This section also presents some case studies for when automating a process could be very helpful. Finally, we will end the book with details on topics to learn further and a path to diving deeper into binary analysis and mobile application reverse engineering.

This part of the book comprises the following chapters:

# 6

# Open Source and Commercial Reverse Engineering Tools

In the chapters so far, we have discussed reverse engineering iOS and Android applications. We mainly used open source tools, with one exception – Hopper Disassembler. Once you start reverse engineering real-world mobile applications, on both Android and iOS, you might find some alternative tools as well. Those alternatives can also be open source or commercial (closed source).

All the open source tools we have used so far, such as APKTool, Ghidra, and Radare2, are all free to use, and the closed source one, **Hopper Disassembler**, is commercially available. So, for the remaining part of the chapter, let's consider what we mean by closed source when we are talking about commercially available tools.

Different reverse engineering tools offer different (and unique) features. It also often becomes a personal choice of which tool you want to use, and for what purpose. For example, I prefer using Hopper Disassembler for all iOS binary reverse engineering over Ghidra. This is primarily because I feel more comfortable working with Hopper than Ghidra. Another reason for someone to prefer Radare2 over Ghidra could be that they like working on a command-line interface rather than a graphical user interface, or the visual mode of Radare2 looks cleaner than having all the disassembled code in one window.

There is a lot of similarity between the features and support that these tools offer, but there are crucial differences, which might make one tool more suitable for a particular job over another.

To help you understand the pros and cons of one tool over another, in this chapter, we will be covering the following topics:

- Some common open source tools for reverse engineering
- Some common commercial (closed source) tools for reverse engineering
- A case study for reverse engineering and the required capabilities of a reverse engineering tool

# Technical requirements

We will be using the Ubuntu virtual machine setup that we used in the previous chapter. This chapter does not have any additional technical requirements.

# Tools for mobile application reverse engineering

There are several options when it comes to choosing a reverse engineering tool for mobile applications. Mobile applications have dex or Mach-O (OS-x) binary files, which are generally the prime focus during reverse engineering. A tool that supports mobile device architectures and can decompile, disassemble, and patch a mobile application would be the right choice of tool for penetration testing. However, sometimes, there are more complex problems that need to be solved, some advance obfuscation performed, and external library files also need to be reverse engineered. In such cases, the reverse engineering tool would need to have these advanced capabilities.

Let's have a look at some of the commonly used open source and commercially available reverse engineering tools, which can be used for mobile applications.

# Open source mobile application reverse engineering tools

Open source tools have their source code publicly available for others to inspect, modify, and enhance. The following list gives information on some of the common open source tools used for reverse engineering:

- **Ghidra**: Probably the most common and advanced open source and free reverse engineering tool used for mobile application reverse engineering. Some of the key capabilities of Ghidra include assembly, disassembly, and decompilation. Ghidra supports a wide variety of processor instruction sets and executable formats.

  The official GitHub repository can be found here: `https://github.com/NationalSecurityAgency/ghidra`.

- **The Radare2 framework**: A multi-architecture and multi-platform tool, capable of assembling and disassembling executables, which it can also perform binary diffing with graphs and extract information. A more detailed capability list can be found on the official Radare2 page: `https://www.radare.org/r/`.

  The official GitHub repository can be found here: `https://github.com/radareorg/radare2`.

- **JADX (only for Android apps)**: Unlike the iOS binary, it is comparatively easier to reverse engineer Android apps. This is mainly because the `dex` file can easily be converted to Java code. One of the most common tools used for this purpose is JADX. This is basically a Java decompiler; it can decompile Dalvik bytecode to Java classes from APK, `dex`, `aar`, `aab`, and more.

  The official GitHub repository can be found here: `https://github.com/skylot/jadx`.

Next, let's look at commercial tools.

# Commercial mobile application reverse engineering tools

There are a lot of very useful and advanced reverse engineering tools available as commercial tools. These are generally closed source and have different licensing models. Let's have a look at some of the famous, commercial reverse engineering tools.

The following list gives information about some of the commonly used closed source tools for reverse engineering:

- **Hopper**: A reverse engineering tool to disassemble, decompile, and debug iOS (ARM) executables (and multiple other executables/binaries as well). Hopper Disassembler comes in a free version as well, with a time restriction of usage. At the time of writing this chapter, a personal license can be purchased from the official website, at a cost of $99–$129, depending upon the time of the licensing model you chose. Having a cheaper and personal license makes this tool very popular among mobile application penetration testers.

  The official website can be found here: `https://www.hopperapp.com/`

- **IDA**: A widely used reverse engineering tool, with the capabilities of a disassembler and a debugger. The professional version (IDA PRO) is one of the most advanced and capable reverse engineering tools available. There are some other versions of IDA available, such as IDA Home and IDA Free. The main difference between IDA Pro and IDA Home is in supported processors and debuggers. A detailed list of differences between all the versions of IDA can be found on the official website: `https://hex-rays.com/ida-pro/#main-differences-between-ida-editions`.

  IDA Free is a binary code analysis tool with some basic IDA functionalities. IDA comes in three types of licenses – named license, computer license, and floating license. More details on IDA license types can be found here: `https://hex-rays.com/licenses/`.

  The price details of all IDA tools and licenses can also be found on the official website: `https://hex-rays.com/cgi-bin/quote.cgi/products`.

  IDA Pro supports a huge number of processor types and binary formats but comes with a costly license model. This makes it more preferable for advanced reverse engineering needs. The official website can be found here: `https://hex-rays.com/ida-pro/`.

- **Binary Ninja**: An interactive disassembler, decompiler, and binary analysis platform. This is another very popular tool among penetration testers and reverse engineers. It also comes with a demo/trial version, with limitations. The license is available for personal (non-commercial), commercial, and enterprise. More details about the pricing can be found here: `https://binary.ninja/purchase/`.

  The license cost of Binary Ninja sits between Hopper Disassembler and IDA.

  The official website can be found here: `https://binary.ninja/`.

- **JEB Decompiler**: A reverse engineering platform to perform disassembly, decompilation, debugging, and analysis of code and document files, manually or as part of an analysis pipeline. JEB Decompiler also comes in a community version that can be used for non-commercial use. The license comes in three types – JEB Android, JEB Pro, and JEB Floating. More details about the pricing can be found at `https://www.pnfsoftware.com/jeb/buy`.

  The official website can be found here: `https://www.pnfsoftware.com/`.

> **Important Note**
>
> The preceding list of open source and commercial reverse engineering tools is not an exhaustive list. These are just the most commonly used and famous reverse engineering tools available.

When it comes to reverse engineering of mobile applications, the choice of tool depends majorly on the following factors:

- The type of mobile application to be reverse engineered – iOS, Android, or both.

- What is the purpose of reverse engineering? This can be for any of the following reasons:

  - Bypassing one or more security controls in the application

  - Understanding logic behind some specific part of the application

  - Finding security issues in the application related to code quality, the security controls implemented, and so on.

  - Analyzing strings and static content stored inside the application package and application binary

  - Exploit writing

- Is it required to patch the binary and create the app with modified code?

We now know about the commonly used open source and commercial reverse engineering tools. Let's also understand the capabilities required from these tools during a penetration test.

# Case study – reverse engineering during a penetration test

One of the primary reasons for reverse engineering a mobile application during a penetration test is to analyze whether the source code has any sensitive information hardcoded, which can further be used by a malicious actor. Other reasons might be bypassing security controls such as SSL pinning, root/jailbreak detection, and role-based client-side access control. However, depending on the type of application and pentest, you might have to spend more effort in performing a more in-depth analysis of a reverse engineered application.

Let's look at one of the case studies. During the penetration test of a FinTech application, it was noticed that the application sent some critical requests to uniquely generated URL endpoints. These endpoints were unique for every request, and in fact, they were getting generated right before the HTTP(s) request was generated. In order to find the way this application generates these URLs, we could do one of the following:

- Reverse engineer the application to find the logic (or function) of how these URLs are generated.

- Perform runtime instrumentation to analyze the application while it is running. In this method, we inject a piece of code inside the running process of the application and then analyze its behavior. One of the commonly used tools to perform runtime instrumentation is Frida (`https://frida.re/`).

On disassembling the application binary and analyzing the logical flow during that specific step, it was revealed that the application creates an SHA-256 hash of the user's ID and session ID. This hash is then used as a part of the URL, such as the following:

1. User ID: `53e726ce-954f-4291-9968-063521b87483`

2. Session ID: `eyJhbGciOiJIUzI1NiIsInR5cCI6IkpXVCJ9.eyJzdWIiOiJ1 c3JfMTIzIiwiaWF0IjoxNDU4Nzg1Nzk2LCJleHAiOjE0NTg4NzIxOTZ9.- LzT9cobGUNs1Z4JSELQFwSxp5JpT5o6KtMO8ySR-20`

3. SHA-256 hash: `FFDC52D453CF836BAC761D7463B85A7AB6EF4DB 511366A27684734E7154C461A`

4. Final unique URL: `https://[RedactedDomainName]/user/admin/ escalate/ FFDC52D453CF836BAC761D7463B85A7AB6EF4DB511366A 27684734E7154C461A`

5.  Final HTTP(s) request:

```
POST /user/admin/escalate/ FFDC52D453CF836BAC761D
7463B85A7AB6EF4DB511366A27684734E7154C461A HTTP/1.1

Host: ://[RedactedDomainName]

Cookie: session= eyJhbGciOiJIUzI1NiIsInR5cCI6IkpXVCJ9.
eyJzdWIiOiJ1c3JfMTIzIiwiaWF0IjoxNDU4Nzg1Nzk2LCJleHAiOjE0N
Tg4NzIxOTZ9.-LzT9cobGUNslZ4JSELQFwSxp5JpT5o6KtMO8ySR-20

User-Agent: Mozilla/5.0 (Macintosh; Intel Mac OS X 10.15;
rv:96.0) Gecko/20100101 Firefox/96.0

Accept: application/json

Accept-Language: en-US,en;q=0.5

Accept-Encoding: gzip, deflate

Content-Type: application/x-www-form-urlencoded;
charset=UTF-8

X-Requested-With: XMLHttpRequest

Content-Length: 276

Sec-Fetch-Dest: empty

Sec-Fetch-Mode: cors

Sec-Fetch-Site: same-origin

Te: trailers

Connection: close


p_web_site_id=3982358328326&p_language=EN&p_show_form_in_
div=N&p_format=HTML&p_print=JSONP&p_joblocation=WWW&p_
current_host=gdx9tof61hm58kkaz4k0dp3qnht8hx
```

> **Important Note**
>
> The SHA-256 hash is created from the `id=53e726ce-954f-4291-9968-063521b87483;` `sessionId=eyJhbGciOiJIUzI1NiIsInR5cCI6IkpXVCJ9.eyJ zdWIiOiJ1c3JfMTIzIiwiaWF0IjoxNDU4Nzg1Nzk2LCJleHAiO jE0NTg4NzIxOTZ9.-LzT9cobGUNslZ4JSELQFwSxp5JpT5o6Kt MO8ySR-20` string. You can use any hash creator tool to create this hash.

The previous implementation suggests that it was intended to hide the actual URL and only allow the request with a valid hash to access the endpoint. The hash is also calculated on the application server and matched with the hash in the request, and then internally redirected to the correct URL. Once it was confirmed that the actual URL is rather static, a content discovery fuzzing of the endpoint was done. This revealed that the internal URL was in fact exposed publicly also. Here, an assumption was made by the developer as well as the DevOps team that no one will be communicating with the actual URL, and hence, no one really validated whether the actual URL was accessible directly or not.

Solving a challenge such as this requires good disassembling, search, and de-obfuscation features in a reverse engineering tool. Open source tools such as Ghidra and Radare2 can very well be used to solve this.

The reverse engineering capabilities required during a penetration test might differ from that during malware analysis. In a penetration test, reverse engineering is generally done to solve a piece of a puzzle or to explore some functionality. However, during malware analysis, reverse engineering is done to explore every piece of an application and understand all hidden features and code. Let's have a look at a malware analysis case study.

# Case study – reverse engineering during malware analysis

Another field of work that requires more advanced reverse engineering skills is malware analysis. Malware researchers spend days and weeks looking at disassembled and decompiled binaries to deduce the application flow. Let's take another case study.

During the analysis of a malware mobile app, it was noticed that the application somehow modifies its behavior depending on factors such as country, language, and applications installed. For a device in the United States, with the English language, and that had financial/banking apps, the application would try to read messages and the transaction history. However, on a different device in a different country, and with dating apps installed, it would try to inject ads in the traffic of other apps. Such a change in behavior cannot be noticed if the application is only used on one device.

However, a good analysis of the disassembled application binary and its associated libraries revealed this behavior of the application. During this analysis, IDA Pro was used for its features such as reference search, populating current code states to a database, and pseudo code. Other tools, such as Binary ninja, can also provide some of these features.

In this case, basic disassembling and search features might not be enough, and hence, we might have to choose a more advanced and capable reverse engineering tool. Also note that, in this case, the associated library files were also reverse engineered.

There are numerous such cases when a more advanced feature is required in a reverse engineering tool or the tools need to support different types of architectures. Interface choice is also a big reason why some might prefer one tool over another.

# Summary

This chapter talked about some commonly used open source and commercial reverse engineering tools. We also discussed some case studies to understand what type of features and capabilities would be required in a tool to solve the problem. For the majority of tasks done during a penetration test, basic disassembling and debugging are needed, so an open source reverse engineering tool would be enough for such a requirement. However, for more advanced features and capabilities, we would have to go with a commercial reverse engineering tool such as IDA Pro or Hopper. It is also important to feel comfortable with the graphical interface (or visual mode) that each of these tools have. That's another reason why someone prefers one reverse engineering tool over another. For the reverse engineering of mobile applications, the important features/capabilities that the tools must have are the disassembly and assembly of OSx and dex files, decompilation, graphing, patching of the binary files, and string search.

Tools such as Ghidra and Radare2 can very well perform the aforementioned tasks. Another important point is that the Android application binaries can be reverse engineered easily, in comparison to the iOS application binary. This is basically because the dex files can be converted to Java code, using a decompiler such as JADX.

In the next chapter, we will look at some of the ways we can use automated scanners, which can also perform a bit of reverse engineering. This might be useful when you have a huge list of applications to reverse engineer and only want to find some basic things in those apps – for example, I am looking for a specific string in a hundred different applications, or I am interested in only checking for some binary protections on hundreds of mobile app binaries. Such tests can be easily automated using open source tools. We will look more closely into this in the next chapter.

# 7
# Automating the Reverse Engineering Process

During a penetration test or malware analysis, reverse engineering is generally performed on one binary (or application) at a time because the aim of reverse engineering is to analyze a single application. However, there can be cases when you need to quickly analyze a lot of applications for some generic details. For example, you want to find out whether a specific method is being used in any of the applications you are working on, or you want to find out whether a specific string (or strings) is a part of any of the available application binaries.

In such cases, it would be really helpful if you could automate these tasks. A static analysis is often the very first step during a black box penetration test of a mobile application. The static analysis helps to quickly analyze the application based on the reverse engineered code, extract strings, analyze the binary for some basic protections, and can also perform a quick malware analysis. So, let's have a look at how we can automate some part of the static reverse engineering using open source tools as well as some scripting.

In this chapter, we will be covering the following topics:

- Using an open source tool to automatically perform a static analysis of Android and iOS applications
- Understanding a case study to automate a few reverse engineering tasks
- Writing scripts to automatically perform a few tasks on binaries

# Technical requirements

We will be using the Ubuntu virtual machine setup, which we used in the previous chapter. In this chapter, we will be using Docker inside the Ubuntu virtual machine to run Mobile Security Framework.

# Automated static analysis of mobile applications

The first step during a black box penetration test is to gather as much information as possible about the target. In the case of a mobile application penetration test (black box), a static analysis of the application package (**Android Application Package (APK)** or **iOS application archive (IPA)**) is done to get a basic idea about the application, as well as to analyze it for some low-hanging vulnerabilities and missing security controls. Let's have a look at things that a static analysis tool can check on an application:

- Extract details about the application from the application's manifest (for Android) or PLIST (for iOS) files.
- Analyze the binary for protections such as **Automatic Reference Counting (ARC)**, code signing, and **Position Independent Executable (PIE)**.

> **Important Note**
> ARC is used for automatic memory management in iOS apps. This is done by handling the reference count of objects at the time of compilation.
>
> The PIE flag is used in iOS application binaries to protect against **Address Space Layout Randomization (ASLR)** by randomizing the application object's location in the memory for every application restart.

- Reverse engineer the application binary to extract Java code (for Android apps), classes (for iOS apps), strings, and so on

- Analyze the binary for the use of insecure APIs.

These are some of the initial test cases to be performed, and a lot of this can be done by analyzing the reverse engineered binary. The most famous tool used for static analysis of mobile applications is **Mobile Security Framework (MobSF)**. Let's set up MobSF and perform automated reverse engineering on SecureStorage iOS and Android applications.

# MobSF

MobSF is an open source security assessment framework for mobile applications. It supports mobile app binaries (APK, XAPK, IPA, and APPX) along with zipped source. We are going to have a look at the use of MobSF in quickly reverse engineering and analyzing the reverse engineered binaries.

The MobSF GitHub repository can be found here: `https://github.com/MobSF/Mobile-Security-Framework-MobSF`.

## Setting up MobSF

As per the official documentation, we can set up this tool on a Linux virtual machine either by cloning the repository and then running the `setup.sh` file or using a Docker container.

> **Important Note**
> MobSF requires Xcode command-line tools for IPA analysis, which can only work on Mac, Linux, and Docker containers.

Let's set up MobSF using the Docker container and analyze the SecureStorage application. To do so, we first need to install Docker on our Ubuntu virtual machine. The steps to do so can be found on the official Docker website: `https://docs.Docker.com/engine/install/ubuntu/`.

To install Docker Engine, you can follow these steps:

1. To set up the repository, use the following code:

```
# sudo apt-get update
# sudo apt-get install \
    ca-certificates \
    curl \
    gnupg \
    lsb-release
```

2. Next, add Docker's GPG key:

```
# curl -fsSL https://download.Docker.com/linux/ubuntu/
gpg | sudo gpg --dearmor -o /usr/share/keyrings/docker-
archive-keyring.gpg
```

3. Set up the stable repository:

```
#    echo \
  "deb [arch=$(dpkg --print-architecture) signed-by=/
usr/share/keyrings/docker-archive-keyring.gpg] https://
download.docker.com/linux/ubuntu \
  $(lsb_release -cs) stable" | sudo tee /etc/apt/sources.
list.d/docker.list > /dev/null
```

4. Update the apt package index:

```
# sudo apt-get update
```

5. Install Docker Engine:

```
# sudo apt-get install docker-ce docker-ce-cli
containerd.io
```

Once Docker Engine has been installed, you can verify whether it's running properly by using the following command:

```
# sudo docker run hello-world
```

The following screenshot shows the output of the preceding command:

```
mare@ubuntu:~$ sudo docker run hello-world
Unable to find image 'hello-world:latest' locally
latest: Pulling from library/hello-world
2db29710123e: Pull complete
Digest: sha256:97a379f4f88575512824f3b352bc03cd75e239179eea0fecc38e597b2209f49a
Status: Downloaded newer image for hello-world:latest

Hello from Docker!
This message shows that your installation appears to be working correctly.

To generate this message, Docker took the following steps:
 1. The Docker client contacted the Docker daemon.
 2. The Docker daemon pulled the "hello-world" image from the Docker Hub.
    (amd64)
 3. The Docker daemon created a new container from that image which runs the
    executable that produces the output you are currently reading.
 4. The Docker daemon streamed that output to the Docker client, which sent it
    to your terminal.

To try something more ambitious, you can run an Ubuntu container with:
 $ docker run -it ubuntu bash

Share images, automate workflows, and more with a free Docker ID:
 https://hub.docker.com/

For more examples and ideas, visit:
 https://docs.docker.com/get-started/
```

Figure 7.1 – Running the hello-world container on Docker

Once we have the container ready, we can now download a prebuilt Docker container and run it directly.

## Running the MobSF Docker container

A prebuilt Docker container can be found on Docker Hub. To download and run the container, use the following code:

Downloading the container from Docker Hub:

```
# sudo docker pull opensecurity/mobile-security-framework-mobsf
```

Here's the output:

```
mare@ubuntu:~$ sudo docker pull opensecurity/mobile-security-framework-mobsf
Using default tag: latest
latest: Pulling from opensecurity/mobile-security-framework-mobsf
7c3b88808835: Pull complete
d6e88c6ed70e: Pull complete
657c5893e3dc: Pull complete
eed1cc7f6b30: Pull complete
e7134de0c6f9: Pull complete
2164cbb23262: Pull complete
1f1df863f23b: Pull complete
959075f50d3d: Pull complete
ac389c590585: Pull complete
0d72d821f1fd: Pull complete
9ce486268472: Pull complete
a57c8a0b1d24: Pull complete
363e250a3e24: Pull complete
4b644fc222bf: Pull complete
Digest: sha256:883fb6c71307c4c59bbcd07b4a1d6f6aa24712ba82267c9099347175b7b4fc49
Status: Downloaded newer image for opensecurity/mobile-security-framework-mobsf:latest
docker.io/opensecurity/mobile-security-framework-mobsf:latest
```

Figure 7.2 – Downloading the MobSF container

Run the following on the container on port 8000:

```
# sudo docker run -it --rm -p 8000:8000 opensecurity/
mobilesecurity-framework-mobsf:latest
```

The output is shown in the following screenshot:

```
mare@ubuntu:~$ sudo docker run -it --rm -p 8000:8000 opensecurity/mobile-security-framework-mobsf:latest
[INFO] 09/Mar/2022 07:37:11 -

  MobSFv3S


[INFO] 09/Mar/2022 07:37:11 - Mobile Security Framework v3.5.2 Beta
REST API Key: 514968ebb150c21551f12b5dc1468c5e0d8055e218db3178413f23abb5fb5071
[INFO] 09/Mar/2022 07:37:11 - OS: Linux
[INFO] 09/Mar/2022 07:37:12 - Platform: Linux-5.11.0-44-generic-x86_64-with-glibc2.29
[INFO] 09/Mar/2022 07:37:12 - Dist: ubuntu 20.04 Focal Fossa
[INFO] 09/Mar/2022 07:37:12 - MobSF Basic Environment Check
No changes detected
[INFO] 09/Mar/2022 07:37:12 - Checking for Update.
[INFO] 09/Mar/2022 07:37:12 - No updates available.
[INFO] 09/Mar/2022 07:37:13 -

  MobSFv3S


[INFO] 09/Mar/2022 07:37:13 - Mobile Security Framework v3.5.2 Beta
REST API Key: 514968ebb150c21551f12b5dc1468c5e0d8055e218db3178413f23abb5fb5071
[INFO] 09/Mar/2022 07:37:13 - OS: Linux
[INFO] 09/Mar/2022 07:37:13 - Platform: Linux-5.11.0-44-generic-x86_64-with-glibc2.29
[INFO] 09/Mar/2022 07:37:13 - Dist: ubuntu 20.04 Focal Fossa
[INFO] 09/Mar/2022 07:37:13 - MobSF Basic Environment Check
Migrations for 'StaticAnalyzer':
  mobsf/StaticAnalyzer/migrations/0001_initial.py
    - Create model RecentScansDB
    - Create model StaticAnalyzerAndroid
    - Create model StaticAnalyzerIOS
    - Create model StaticAnalyzerWindows
[INFO] 09/Mar/2022 07:37:14 - Checking for Update.
[INFO] 09/Mar/2022 07:37:14 - No updates available.
[INFO] 09/Mar/2022 07:37:15 -
```

Figure 7.3 – Running the MobSF container

Once this is done, you should be able to access the MobSF dashboard at
`http:127.0.0.1:8000`.

Figure 7.4 – The MobSF dashboard

Now that we have the MobSF tool running, we can perform a static analysis of iOS and
Android apps.

# Performing a static scan on SecureStorage

Once we have MobSF running, simply drag and drop IPA or APK to complete the static
analysis. Once the scan is complete, you will be presented with a report.

## Static analysis of APK

During a static analysis of APK, MobSF performs the following tasks:

1. Decompiling and extracting content such as hardcoded certificates/key stores
2. Converting AXML to XML
3. Extracting and analyzing manifest data
4. Creating the Java code
5. Converting DEX to SMALI
6. Extracting strings

As you can see, a lot of these tasks are the same as the ones we performed during our manual reverse engineering of the Android application. Once the analysis is done, we can use the MobSF dashboard to download the reverse engineered JAVA code, strings, `AndroidManifest` file, and more for a manual analysis. MobSF also performs an automated analysis of the extracted content and provides us with a report containing all the information (including any security issues discovered).

## Static analysis of IPA

During the IPA analysis, MobSF performs the following tasks:

1. Extracting the IPA
2. Analyzing the `Info.plist` file
3. A Mach-O analysis of the binary
4. Dumping classes from the binary
5. Binary analysis

Once the analysis is done, we can download the reverse engineered entities, such as `strings`, `classdump`, and `Info.plist`.

MobSF also performs a quick check on the binary for some protections.

| PROTECTION | STATUS | SEVERITY | DESCRIPTION |
| --- | --- | --- | --- |
| ARC | True | info | The binary is compiled with Automatic Reference Counting (ARC) flag. ARC is a compiler feature that provides automatic memory management of Objective-C objects and is an exploit mitigation mechanism against memory corruption vulnerabilities. |
| CODE SIGNATURE | True | info | This binary has a code signature. |
| ENCRYPTED | False | warning | This binary is not encrypted. |
| NX | True | info | The binary has NX bit set. This marks a memory page non-executable making attacker injected shellcode non-executable. |
| PIE | True | info | The binary is build with -fPIC flag which enables Position independent code. This makes Return Oriented Programming (ROP) attacks much more difficult to execute reliably. |
| RPATH | True | warning | The binary has Runpath Search Path (@rpath) set. In certain cases an attacker can abuse this feature to run arbitrary executable for code execution and privilege escalation. Remove the compiler option -rpath to remove @rpath. |
| STACK CANARY | False | high | This binary does not have a stack canary value added to the stack. Stack canaries are used to detect and prevent exploits from overwriting return address. Use the option -fstack-protector-all to enable stack canaries. |
| SYMBOLS STRIPPED | False | warning | Symbols are available. To strip debugging symbols, set Strip Debug Symbols During Copy to YES, Deployment Postprocessing to YES, and Strip Linked Product to YES in project's build settings. |

Figure 7.5 – IPA binary analysis

Note that this is the same information that we collected during the manual reverse engineering of the SecureStorage binary using Radare2 in *Chapter 5, Reverse Engineering an iOS Application (Developed Using Swift)*.

The REST APIs in MobSF can also be used to further automate the process of static analysis of the applications.

Using MobSF, we have automated some part of the basic reverse engineering that we did manually in *Chapter 3, Reverse Engineering an Android Application*, and *Chapter 4, Reverse Engineering an iOS Application*. However, this is in no way an alternative to a deep manual analysis of the reverse engineered binary implemented during a penetration test or malware analysis.

However, using a tool such as MobSF can be extremely helpful while performing an analysis of a large number (>5) of applications in a quick timeframe. The automation reduces the overhead time.

Let's understand this using a case study.

# Case study one – automating reverse engineering tasks

During a research project, we need to analyze how secure modern mobile applications are and what percentage of these applications do not follow some best security practices of binary protection, such as a stack canary and a PIE flag.

In order to complete such research on a wide range of IPAs, we would need to automate the process of binary analysis and reverse engineering. This is where using a tool such as MobSF can be very productive. Here is how we performed such checks on more than 500 applications:

1.  We stored all IPAs at one location.

2.  We then used the MobSF REST APIs to automate the static analysis of binaries one by one:

    - By uploading the file: `api/v1/upload`

    - By scanning the uploaded file: `/api/v1/scan`

3.  Once the analysis is done, a JSON format of the report could be fetched and analyzed to find the value of checks we are interested in:

    • By generating the JSON report: `api/v1/report_json`

    We then grepped the JSON report for our interesting values.

```
mare@ubuntu:~$ curl -X POST --url http://localhost:8000/api/v1/report_json --data "hash=24c5ca219920251169bc01a
7017d9bad" -H "X-Mobsf-Api-Key:514968ebb150c21551f12b5dc1468c5e0d8055e218db3178413f23abb5fb5071" --output repor
t.json
  % Total    % Received % Xferd  Average Speed   Time    Time     Time  Current
                                 Dload  Upload   Total   Spent    Left  Speed
100 37748    0 37711  100    37   171k    172 --:--:-- --:--:-- --:--:--  170k
mare@ubuntu:~$ █
```

Figure 7.6 – Extracting the JSON report

Once we have the report in JSON format, we can quickly hunt for specific data that we are looking for, such as the status of binary protection. There can be a lot of other scenarios where you might need to perform some part of reverse engineering on a huge number of applications (or binaries). If the test case falls under the list of automated scanners such as MobSF, then you can use it. However, sometimes, the test case is very specific, and you can't run a static scan and compare reports.

# Case study two – automating test cases to find security issues

During an audit, we noticed that all mobile applications developed by a specific team used a list of common secrets and hardcoded values in the code. As it was also a black box penetration test, we did not have any source code but had a list of 10+ Android applications to test. We wanted to find out how many of these applications have the same secrets and hardcoded accounts inside the application code. One way of doing this could have been by manually extracting strings from each of these application binaries and then searching for them. But we automated this part a little bit by following these steps:

1.  Extracting all `dex` files from the APKs, using the `unzip` utility

2.  Running strings on all `dex` files and saving the result in different text files

3.  Grepping through all the text files containing strings to search for our specific strings

A simple script to automate this would look like this:

```
#!/bin/bash
#For all files in the directory (all APKs are in this
directory):
```

```
for file in *.apk
do
    echo "Extracting $file" #Printing the name of file
    mkdir classes."$file" #Creating a directory with app name
    unzip "$file" -d classes."$file"/ '*.dex' #extracting the
content of APK in temp directory


    #mv temp/classes* "classes.$file" #Moving all dex files to
the new directory
    #Running string on each dex file and saving the output to a
single strings.txt file


    find classes."$file"/ -iname '*.dex' -exec strings "{}"
>classes."$file"/strings.txt \;
done
```

> **Important Note**
>
> The preceding script might also create directories with the names `classes.`
> `run.sh` (`run.sh` is the bash script in the same directory) and `classes.`
> `temp` (`temp` is the folder being created). To remove these directories, add the
> following two lines in the bash script:
>
> ```
> rm -rf  classes.run.sh
> rm -rf classes.temp
> ```

The preceding script will automate the process of extracting dex files from a group of APKs and will then extract strings from all dex files from the same APK and save them in the strings.txt file (in each APK directory). Once we have the strings for all APKs, the bash script can be extended to search for a specific string in those sets of strings. txt files.

This is another case study of how you can automate some part of the reverse engineering process using a custom bash script.

# Summary

This chapter talked about some case studies and gave some examples when automating the reverse engineering process that might be helpful. Remember that reverse engineering can be extremely in depth depending on what you are trying to achieve with it. Finding hardcoded strings, class names, and more are the simpler tasks done through reverse engineering. However, there can be a lot of complex challenges for which an in-depth, manual analysis and reverse engineering might be required.

Automating all such requirements is not always possible. But it is usually a good idea to automate the part of your work that you will need to perform again and again – for example, extracting strings and classes from mobile application binaries.

That's it for this chapter. In the next and final chapter of this book, we will summarize what we have discussed, what more can be explored in order to enhance your knowledge, and what should be the way ahead.

# 8
# Conclusion

Software reverse engineering, in simple terms, is the art of taking apart an application or software to understand its internal workings. The way a piece of software/code functions depends upon several factors such as the programming language, the CPU architecture it is built for, and programming practices. The process of reverse engineering as well as analyzing the reverse engineered software, depends on the type of architecture it was developed for, the type of programming language, and so on.

For mobile application reverse engineering, the initial phase requires an understanding of an application package structure, how it is developed, the programming language, binary format, the application package type, and so on. With this knowledge, we start the process of reverse engineering; in the case of an Android app, we use JADX because we know that Java code can be extracted from dex files. However, for an iOS app, we disassemble the binary using a disassembler tool such as Ghidra.

To explore deeper inside the reverse engineered binary, we would need to have a better understanding of assembly language, architecture, native libraries, other programming languages, and so on.

This book focuses on teaching how to get started with reverse engineering iOS and Android applications. But this is just the tip of the iceberg. So, let's discuss what more there is to learn, and how to get more expertise in the field of Android and iOS application reverse engineering.

# Excelling in Android application reverse engineering – the way forward

This book has covered the very basics of getting started with understanding how an Android application is developed, its internals, and how to reverse engineer an **Android Package** (**APK**).

Depending upon the objective you are trying to achieve with reverse engineering, sometimes, only decompiling the APK will give you the answer, but sometimes, you might have to go steps ahead and analyze the Smali or Java code.

Android applications can also contain native libraries. These are the code compiled for a specific architecture, mainly for intensive tasks. From a malware analysis point of view, it becomes critical to reverse engineer the native libraries as well because malicious code could also have been hidden inside these libraries. But to be able to reverse engineer the compiled code, one would need to have an understanding of assembly language, ARM and x86 architecture details, and so on. Here are a few things to learn if you want to excel in the art of reverse engineering Android applications:

- Learn the ARM assembly
- Learn more about binary reverse engineering
- Learn about Android native libraries
- Learn about the dynamic loading of Dalvik Executables in Android apps
- Learn about runtime instrumentation using tools such as Frida

The preceding list is provided to help you see the way ahead in learning more about Android application reverse engineering. A lot of learning eventually comes with practice, so as you work on reverse engineering different applications, you will eventually learn more binary reverse engineering skills, how developers usually write the program flow, and so on.

# Excelling in iOS application reverse engineering – the way forward

This book has covered details such as IPA architecture, contents, binary formats, disassembling the iOS application binary, navigating through the disassembled code, and so on. As we know, iOS applications do not have Java bytecode but instead have the compiled binary. So, the only way to understand the functioning of an iOS application is through understanding the disassembled binary. Excelling in iOS application reverse engineering also requires a lot of other skills than those that have been discussed in this book.

Here are a few things to learn if you want to excel in the art of reverse engineering iOS applications:

- Learn more about binary reverse engineering
- Learn about Mach-O and dynamically loaded code
- Learn about runtime instrumentation using tools such as Frida

# Utilizing reverse engineering skills

As a mobile application penetration tester, malware researcher, exploit writer, and so on, you will often be required to reverse engineer different types of applications. As discussed in the previous chapters of this book, mobile application reverse engineering can be helpful in a lot of ways:

- Bypassing security controls such as certificate pinning and root/jailbreak detection
- Analyzing the application flow and bypassing any runtime security control, such as input encryption
- Analyzing mobile malware applications
- Performing security assessment of an application
- Finding security issues in code or an application package
- Modifying the behavior of an application and repacking it

In more than 500+ black-box mobile application penetration tests that I have performed so far, reverse engineering is the first step. It gives a great insight into how an application is created, some basic details, and low-hanging vulnerabilities (if any) – for example, *hardcoded sensitive details* in application code is a very common security issue discovered during penetration tests. To find such a security issue, I have a list of common keywords that can be searched in the extracted strings from the application binary. Often, reverse engineering provides great insights into some hidden functionalities as well. Let's have a look at one such case study.

# Exposing unreleased features in an application through reverse engineering

Mobile applications are getting new updates almost every day now. Developers need to keep up with the speed of new releases, and to do so, production mobile apps often also have code for some unreleased features. A very common feature is a hidden debug menu or a hidden configuration/analytics menu.

While analyzing a reverse engineered mobile application, it was noticed that a major part of code was available that did not have any UI for normal users. However, on tapping a few times on the logo on the **About us** screen, a hidden screen is shown, which asks for a password.

This hidden screen gives access to some debug/internal config features of the application, enabling you to disable the license check, bypass all access control checks, and so on. This can be a goldmine of security issues if found in a production application. The only way to discover these types of features is through reverse engineering and analyzing the disassembled binary or decompiled Java code.

As discussed in *Chapter 7, Automating the Reverse Engineering Process*, automating some parts of reverse engineering is also very useful, especially when a review is to be done on a huge number of applications.

# Summary

Being able to reverse engineer an application and understand (as well as modify) its internal working is a great superpower to have. We discussed how mobile applications are developed, the basics of reverse engineering mobile applications, how reverse engineering can be useful during security assessment, as well as malware analysis, and how to automate some parts of reverse engineering.

The path to learning does not stop here, as this is just the start of what is possible with reverse engineering. With this, I will end this book here, and I hope you all will have a lot of fun reverse engineering mobile applications to find and fix security issues.

# Index

Subscribe to our online digital library for full access to over 7,000 books and videos, as well as industry leading tools to help you plan your personal development and advance your career. For more information, please visit our website.

## Why subscribe?

- Spend less time learning and more time coding with practical eBooks and Videos from over 4,000 industry professionals

- Improve your learning with Skill Plans built especially for you

- Get a free eBook or video every month

- Fully searchable for easy access to vital information

- Copy and paste, print, and bookmark content

Did you know that Packt offers eBook versions of every book published, with PDF and ePub files available? You can upgrade to the eBook version at packt.com and as a print book customer, you are entitled to a discount on the eBook copy. Get in touch with us at customercare@packtpub.com for more details.

At www.packt.com, you can also read a collection of free technical articles, sign up for a range of free newsletters, and receive exclusive discounts and offers on Packt books and eBooks.

# Other Books You May Enjoy

If you enjoyed this book, you may be interested in these other books by Packt:

**Ghidra Software Reverse Engineering for Beginners**

A. P. David

ISBN: 9781800207974

- Get to grips with using Ghidra's features, plug-ins, and extensions
- Understand how you can contribute to Ghidra
- Focus on reverse engineering malware and perform binary auditing
- Automate reverse engineering tasks with Ghidra plug-ins
- Become well-versed with developing your own Ghidra extensions, scripts, and features
- Automate the task of looking for vulnerabilities in executable binaries using Ghidra scripting
- Find out how to use Ghidra in the headless mode

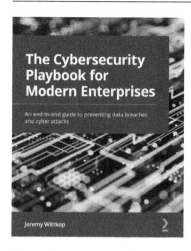

**The Cybersecurity Playbook for Modern Enterprises**

Jeremy Wittkop

ISBN: 9781803248639

- Understand the macro-implications of cyber attacks
- Identify malicious users and prevent harm to your organization
- Find out how ransomware attacks take place
- Work with emerging techniques for improving security profiles
- Explore identity and access management and endpoint security
- Get to grips with building advanced automation models
- Build effective training programs to protect against hacking techniques
- Discover best practices to help you and your family stay safe online

# Packt is searching for authors like you

If you're interested in becoming an author for Packt, please visit `authors.packtpub.com` and apply today. We have worked with thousands of developers and tech professionals, just like you, to help them share their insight with the global tech community. You can make a general application, apply for a specific hot topic that we are recruiting an author for, or submit your own idea.

# Share Your Thoughts

Now you've finished *Mobile App Reverse Engineering*, we'd love to hear your thoughts! Scan the QR code below to go straight to the Amazon review page for this book and share your feedback or leave a review on the site that you purchased it from.

`https://packt.link/r/1801073392`

Your review is important to us and the tech community and will help us make sure we're delivering excellent quality content.

www.ingramcontent.com/pod-product-compliance
Lightning Source LLC
Chambersburg PA
CBHW060141060326
40690CB00018B/3936